CARTOGRAPHIC ENCOUNTERS

Cartographic Encounters

Indigenous Peoples and the Exploration of the New World

John Rennie Short

REAKTION BOOKS

Published by Reaktion Books Ltd
33 Great Sutton Street
London EC1V 0DX, UK

www.reaktionbooks.co.uk

First published 2009

Printed and bound in Great Britain
by Cromwell Press Group, Trowbridge, Wiltshire

British Library Cataloguing in Publication Data
Short, John R.
Cartographic encounters : indigenous peoples and the exploration of the New World
1. Cartography – West (U.S.) – History – 19th century
2. Indian cartography – West (U.S.) – History – 19th century
3. Indians of North America – West (U.S.) – First contact with Europeans
4. West (U.S.) – Discovery and exploration
I. Title
978'.02

ISBN: 978 1 86189 436 6

Contents

PART I | introduction

Early America was a cloth woven from many threads,
but the Indian strands that ran through it have often been ignored,
forgotten, and allowed to fade from the nation's history.

Colin Calloway, 1997

1 | Creation Myths and Cartographic Encounters

There is a timeline printed in the official *Montreal Official Tourist Guide* (2006–7); it begins with an entry beside the date 1535, 'Jacques Cartier, who discovered Canada, returns upriver to the island that will later bear the name Montreal.' It goes on, 'Impressed with the mountain, he climbs it and calls it Mont Royal.'[1]

It is a story that explains and signifies a city's birth in history. A similar tale is told for many cities and places around what they called the New World; heroic exploration by a founding father (or fathers) followed by an act of naming that marks a beginning in time and an origin in space. It is the dominant creation myth forged by Europeans for their New World.

In this book I want to look more closely at this and related myths. I will argue that their focus on discovery and exploration ignores the large, indispensable and vital roles played by the indigenous people. The tales should be less about discovery and more about collaboration. I want to replace these persistent myths with the notion of cartographic encounters. My basic argument is that the successful European and later American exploration of the New World resulted from the exchange of information between newcomers and indigenous people. This exchange can be read from contemporary accounts and is embodied in contemporary maps. I will deconstruct the texts and the maps to reveal that the

1 *Across the Continent: 'Westward the Course of Empire Takes its Way'*, an 1868 print by Frances Flora Bond Palmer published by Currier & Ives.

European/American depiction of the land was very dependent on indigenous peoples.

Wider debates

Three wider debates situate the ideas in this book. The first is the changing account of indigenous people in the history of settler societies such as the United States. Initially the natives were noted, if at all, as background to the triumph of European exploration, colonization and settlement of the continent. They were portrayed as a barrier to European colonization, as part of the 'natural' landscape or simply passed over in silence. We have a visual representation of this approach in the 1868 painting and subsequent print by Frances Flora Bond Palmer entitled *Across the Continent: 'Westward the Course of Empire Takes its Way'*. Reproduced many times, Palmer's work was very popular and widely circulated in the United States. The title is taken from the sixth stanza of Anglican bishop and philosopher George Berkeley's 1726 poem 'On the Prospect of Planting Arts and Learning in America', in which he presents the New World of North America as the ultimate site of civilization's progress. Illus. 1 is the 1868 Currier & Ives lithograph based upon the painting. The image shows men clearing the forest around a small settlement and a rail line tracking across a flat plain, heading westward into the distance and the future. *Across the Continent* extols the idea and describes a practice of US progress. The church and school embody acts of civilization. If you look carefully you can see a couple of Native Americans on horseback, separate from the zone of civilization, partly hidden and probably choking on the smoke and steam from the train's engine. Blocked and obscured, the original inhabitants of the continent are marginal to the optimistic forward push west towards the beckoning horizon, reduced to bit players, representatives of a dying past.

Countless western movies and novels, as well as serious academic studies, give Native Americans supporting roles, passive roles or no roles, passing them over in silent condescension. Things began to change in the second half of the twentieth century. As US progress began to seem at best a mixed blessing and for some a downright regression, Native Americans were recast in a rather romantic light, living in the pristine, pre-Fall Eden before the encroachment of urbanization, industrialization and commercialization. The career of the great director of the western John Ford (1894–1973) signals the shift. His earliest box-office success was *The Iron Horse*, which opened in 1924. It was a celebration of Manifest Destiny, the conquest of the West, the defeat of the Indians and the construction of the continental railway, the filmic equivalent of Palmer's painting. Later, in *Stagecoach* (1939), he portrays the Apaches as savages who kill settlers and rape women. Their attack on the stagecoach of the movie title is almost successful but for the US cavalry who arrive in the nick of time to save the day. The Native Americans of the early Ford westerns are a threat and a nuisance, an irritant in the heroic making of the United States. In his later movies, especially *Fort Apache* (1948), *She Wore a Yellow Ribbon* (1949), *Wagon Master* (1950) and *The Searchers* (1956), Native Americans assume a more ambiguous role. Their defeat is still necessary for the making of the United States, but is now as much a source of regret as a cause for celebration. And in *Cheyenne Autumn* (1964) the camera turns a full 180 degrees to show the perspective of Native Americans displaced from their land. Ford remarked that he wanted to show the other point of view for a change. *Cheyenne Autumn* is a forerunner of later pro-Native American westerns such as *Soldier Blue* (1970) and *Dances with Wolves* (1990), in which the savages are the US cavalry.[2]

Francis Jennings provides an academic version of *Cheyenne Autumn* in his 1975 book *The Invasion of America*, in which the story of the European discovery, exploration and settlement of North America is represented as a story of genocide; it is history from the other side of the frontier, an account of loss and dispossession, of death and destruction. Jennings's argument, often shorn of its subtlety, soon became a standard depiction in this more critical assessment of national history and national identity: it changed the Native Americans from shadows to victims, from savages to heroes. Richard White modified this stereotype somewhat in his 1991 book *The Middle Ground*, which showed the frontier not only as a place of defeat but also of negotiation. Contemporary western historical writing is now more aware of Native Americans as active agents as much as victims and of Europeans as constrained as well as triumphant. We now have a vigorous school of history that shows Native Americans as 'friend and foe, trader and neighbor, fellow diplomat and fellow Christian'.[3]

However, I feel that perhaps we have moved too far in the direction of seeing this interaction as one of compromise and negotiation. White's sensitive analysis, like Jennings's before, has been reduced to a simplistic conclusion, the persistence of middle ground. But there was no lasting middle ground: one society became dominant at the expense of the other; one culture won, and one lost. I will use the term *symbiotic destruction* to refer to the European–Native American interaction, a relationship that involved choices and constraints, compromises and negotiations as well as conflicts and struggles, limitations on Europeans and exercises of Native American power, but set within the long-term story of eventual European victory and Native American defeat.

The second debate that helps to frame this book is the recent radical shift in the history of cartography. A story of mapmaking's increasing scientific rationality has long dominated the narrative. Maps were milestones along this journey, laid out along an incline of an increasing knowledge of the world. This whole discourse was undermined by the postmodern turn, which emphasized maps as social constructions, stories marked by purposeful erasures and silences as well as inscriptions and disclosures. Influential scholars such as Brian Harley deconstructed maps for their ideology, their political undertones and their social contexts. Maps are now no longer seen as uncomplicated pictures of the geographic world: they are now understood to reflect power relations and embody the knowledge and ignorance, articulations and silences of the wider social world. Map accuracy and provenance are no longer the only considerations in this new history of cartography: it is now important to uncover maps' narrative context, their truths as well as their lies, and to see the act of mapping as a political act as much as a scientific practice.[4]

There is a third debate nestled within this postmodern history of cartography. Malcolm Lewis writes of cartographic encounters between Europeans, European–Americans and Native Americans. He defines these encounters narrowly in terms of maps made by Native Americans in the course of their encounters with whites. I use an even wider definition of cartographic encounters, which recognizes that the maps made by white explorers drew heavily upon an indigenous cartographic contribution. Cartographic encounters result in maps made by the white Europeans in contact with the Native Americans as well as maps made by the indigenous people in contact with the white Europeans.[5]

A frontier is an important site for cartographic collaborations that embody a *symbiotically destructive* relationship. This was a relationship that allowed Native Americans to parlay their deeper and wider knowledge of the land into a strong bargaining position but, in this very exchange, where they had some leverage, lay the roots of their ultimate loss of land. They used the resources they had, such as geographical knowledge, to gain short-term advantages, such as trade goods and alliances with the powerful newcomers, but over the long term these cartographic encounters gave the newcomers enough knowledge and power to render superfluous the indigenous people. These encounters encapsulate the shared endeavour, the negotiated compromise that contains the seeds of defeat and victory.

Cartier and Columbus

Let us return to Jacques Cartier. The main facts of his explorations are the following: very little is known of his early life, but we do know that he undertook sea voyages to Brazil and Newfoundland before the Bishop of Saint-Malo proposed his name to the king of France as a likely leader of an expedition to the New World to find mineral wealth and a westward passage. The king accepted his nomination, and in April 1534 Cartier set off as leader of two ships and 61 men. He sailed to what is now Canada and entered the Gulf of St Lawrence, returning to France in September without any gold. He sailed again in 1534, this time with three ships, and sailed all the way up the St Lawrence River, landing at the site of Montreal, thus ensuring his later entry into the *Tourist Guide*. He spent two years in Canada before returning to France with news of a great river route, and returned to Canada for a third and final time in 1541. On his return to Saint-Malo he became a prosperous businessman and died on 1 September 1557. Cartier's explorations were soon incorporated into other maps. Champlain's explorations and maps of 1613 and 1632 are based on Cartier (illus. 2).

Cartier's name lives on. In Montreal alone a bridge, a basin, a street, a place and a pier bear his name (see illus. 3). There are Jacques Cartier parks in Ottawa, Prince Edward Island and New York State. His name also lives on as the 'explorer' who 'discovered'

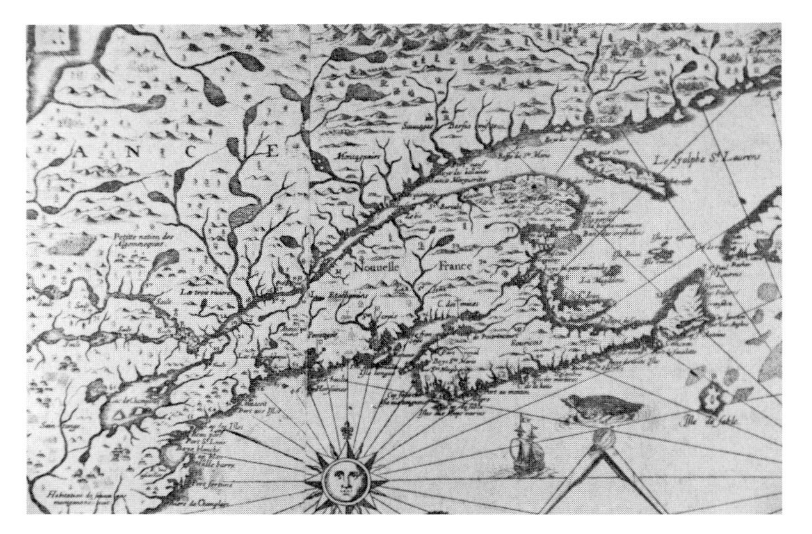

2 Samuel de Champlain's 1613 map of New France, from *Les Voyages du sieur de Champlain Xaintongeois* (Paris, 1613).

the region. In the 1992 *Atlas of North American Exploration*, the introduction to a two-page spread on his voyages summarizes his contribution thus: 'The Breton seafarer Jacques Cartier finds one of North America's great rivers and travels a thousand miles into the interior of the continent.'[6]

A record of Cartier's journeys, purportedly in his own words, exists. And when we look at these words in some detail, a more complicated picture emerges than the simple story implied in the *Atlas* entry or the official city guide timeline.[7] By mid June 1534 Cartier and his two ships entered the Gulf of St Lawrence. They sailed south along the western coastline of Newfoundland and across the gulf to what is now called the Gaspé Peninsula. On 24 July, 'We had a cross made at thirty feet high, which was put together in the presence of a number of savages.'[8] The 'savages' were Iroquoians from further south, on their annual summer fishing expedition. The leader of the tribe, Donnacona, his brother and three sons paddled up in canoes and disputed the implied claim. 'And pointing to the cross he made us a long harangue . . . as if he wished to say that all this region belonged to him, and that we ought not to have set up this cross without his permission.'[9]

The French, by pretending they will trade an axe for the cloth the chief is wearing, lure the Native Americans close to the side of the ship, seize the canoe, capture the Iroquoians and take them on board the French ship. They kidnap the chief's two sons Taignoagny and Dom Agaya, who are used as guides for the rest of the voyage and taken back to France. During that winter in Saint-Malo they provide Cartier with much information about eastern North America. From this forced collaboration Cartier obtains geographic information that enables him to situate his first voyage and plan his second. He 'discovers' the St Lawrence because his captives guided him up the St Lawrence on his second voyage. '[T]he two savages assured us that this was the way to the mouth of the great river of Hochelaga [St Lawrence] and the route towards Canada . . . and also that farther up, the water became fresh.'[10]

The use of native informants continues throughout Cartier's travels,

> And we learned from Donnacona, from Taignoagny, Dom Agaya and the others that . . . the river of the Saguenay reaches to the Saguenay, which lies more than a moon's journey from its mouth, towards the west-northwest; but that after eight or nine days' journey this river is only navigable from small boats; that the regular and direct route to the Saguenay, and the safer one, is by the river [St Lawrence] to a point above Hochelaga.[11]

In May 1536 Cartier kidnaps the chief, his two sons and two others and takes them back to France. They all die there, never able to return to

3 Street sign in Montreal.

their homeland. This is a more complicated story than the condensed version of Cartier sailing up the St Lawrence and 'discovering' Montreal. It is a tragic tale, with Cartier's 'discovery' predicated upon Native American knowledge and Native American informants.

While Cartier's story is unique, it is of a piece with the story of the early European discovery of the New World. To read Christopher Columbus, for example, is to read of continual use of native informants. Here, in a letter to the king and queen of Spain, he describes his travels in his fourth voyage to the New World along the Mosquito Coast of present-day Nicaragua:

> I reached the land of Cariay, where I stayed to repair the ships and rest the crew . . . Two Indians brought me to Carambaru . . . They gave me the names of many places on the sea-coast where they said there was gold and goldfields too . . . In all these places, I had visited, I found the information given me true.[12]

A few weeks later in October 1502 at La Huerta:

> When about to sail, Columbus seized seven of the people, two of whom apparently the most intelligent, he selected to serve as guides . . . The inhabitants of Cariari manifested unusual sensibility at the seizure of their countrymen. They thronged the shore and sent off four of their principal men with presents to the ship, imploring the release of the prisoners. The admiral (Columbus) assured them that he only took their companions as guides for a short distance along the coast and would restore them soon in safety to their homes, he ordered various presents to be given to the ambassadors; but neither the promises nor the gifts could soothe the grief and apprehension of the natives beholding their friends carried away by beings of whom they had such mysterious apprehensions.[13]

In the cartographic encounters between the 'explorers' and the indigenous people, the land was mapped, narrated and claimed. But it was not a case of explorers working on their own, plotting their way through a foreign land. They were helped, willingly and often unwillingly, as in the case of Cartier and Columbus, by the native peoples.

The Cartier example also encapsulates both the memorialization and the amnesia of history. Cartier is remembered, celebrated, his name lives on in the named places. There is a sturdy quality to this particular myth. The name of a 'French explorer' is kept alive long after a French possession becomes part of the British Empire and then part of an independent country. Historical memory is selective: there are few memorials to Donnacona, Taignoagny or Dom Agaya. Their names largely have been forgotten. Ask most Canadians, and they will know about Cartier. Mention the names of Donnacona, Taignoagny or Dom Agaya and note how few know who they were or what role they played in the 'discovery' of the St Lawrence.

Method

In this book I want to excavate the names of the other Donnaconas, Taignoagnys and Dom Agayas involved in the exploration and mapping of the New World. I will use the explorers' journals. I will scour their own narratives and those of their companions for evidence of cartographic encounters. The strategy is one of necessity. We have few other written records. These accounts often are written in the form of diaries with events set down in chronological order: they often have the directness of confessionals, but this is as much manufactured as authentic. Most of them were written after the fact, sometimes by other people, but always with an eye to making the explorer look good. The accounts have a selective quality, playing a double role of describing events while justifying actions. They suppress as much as they disclose, but this bias ensures that when they do reveal evidence of cartographic encounters, it is obviously of some importance. As in the case of Cartier, we can use his own words to discover the part played by others. With the benefit of hindsight and a sensitive postmodern ear, we can detect the contribution of the Native Americans embedded in the explorers' own journals and narratives. We now have the ability to detect the perspective of the native others. One example: when Cartier returns to the region in 1535, he releases Taignoagny and Dom Agaya to the chief. A week later they come across the tribe:

> All came over towards our ships . . . except the two men we had brought with us . . . who were altogether changed in their attitude

> and goodwill, and refused to come on board our ships, although many times begged to do so. At this we began somewhat to distrust them.[14]

When I first read this I burst out laughing. You kidnap two men, take them to the other side of the world, and then you are surprised that they will not come back on board your ship! And you distrust them! We read the journals with a very different perspective to that held by the writers and the early readers of these texts.

Reliance on the journals precludes the use of the more sparse narratives. We can make little, for instance, from *Warre's Sketches on North America and the Oregon Territory*. In a model of English understatement he describes a journey that started on 5 May 1845 in Montreal. With fifteen Canadiens and Indians in two large canoes, he travelled across the continent reaching the Pacific on 25 August. The book is all of five pages long including a map of the journey. Descriptions are brief and the whole undertaking is summed up in little more than a throat clearing. We can learn little from such laconic narratives. Thankfully most of the narratives are wordy in the extreme and provide a much more substantive body of text.[15]

Most of the narratives are complex forms of representation that describe four spaces. There is the *empty space* that awaits the full unfolding of the colonial/imperial project. Even when the narratives contain descriptions of the indigenous people, the land is conceptualized as a blank page for colonial/imperial expansion. Then there is the *occupied space* of an inhabited land with a due recognition of a humanized landscape full of people. The consequences and implications of a still settled space and its tensions with an empty space invoke many responses from the geopolitical to the moral. Then there is the *travelled space* in a reportage that recounts events, describes scenes and personal responses. Finally, there is the *scientific space* of measurement, theorizing and empiricizing that figures more largely as the nineteenth century unfolds.

The dominant form of space represents a shift over time. In the late eighteenth century William Bartram journeyed through the southern states. In a four-year journey that started in 1773 he moved through the southern colonies recording flora, fauna and the lives of the indigenous peoples. His narrative was published in 1791; it was entitled *Travels*. The word is important. To travel across land is to move through an occupied space. By the nineteenth century many of the western journeys became explorations, voyages of discovery. The shift from travels to expeditions/explorations is a shift in the dominant form of spatial representation; it is a change from seeing an occupied space to an empty space. Moving from travels to explorations signifies a whole different way of seeing the world and denotes the change from a passage through a humanized landscape to a journey through the empty space of colonial sentiments and imperial projections. In this book I want to explore the tensions between the depiction of an empty space and the practice of moving through an occupied space.

Colin Calloway has written a scholarly yet accessible book that elaborates on the theme of native–white interactions as an important element in the making of early America. His work forces us to think of cultural exchanges rather than of fixed categories. Symbolic and material goods as well as information and intelligence were appropriated as well as traded. He reminds us, 'Early America was a cloth woven from many threads, but the Indian strands that ran through it have often been ignored, forgotten, and allowed to fade from the nation's history.' In this book I want to highlight the Indian warp in the fabric that is the exploration and mapping of the territory.[16]

2 | Amerindian Mappings

In their encounters European explorers drew upon the geographic knowledge, spatial sensitivity and cartographic abilities of the indigenous people. The traditional history of cartography tends to discount the cartography of indigenous people in the mapping of the New World, but more recent studies are much more generous. One example of this shift in sensitivity is the transformation of the classic work, *The Southeast in Early Maps* by William P. Cumming, first published in 1958. A study of the maps and mapmaking of the southeast of North America before the American Revolution, it barely recognized the indigenous contribution. Cumming's work is a typical study of New World mapping as a white European practice. But in the third edition published in 1998, enlarged and revised by Louis De Vorsey, Jr, there is a specific chapter on the role of American Indians in the early mapping of the southeast as well as a deeper appreciation of their contribution to maps previously considered as all European. We are now much more aware of the mapmaking abilities of indigenous people in the New World both in the pre-Columbian age and the post-Columbian encounter.[1]

4 Hernando De Soto's map of the Gulf of Mexico, *c.* 1544.

Cartography in the Pre-Columbian New World

Cartographic practices varied. Indigenous cartographers drew upon a wide variety of media to represent space, place and territory. A diversity of mapmaking practices occurred throughout the continent. In the West maps were carved on the basalt surface of the aptly named Map Rock in Idaho. These maps represent an entire river basin covering almost 32,000 square miles. In the far North the Inuit have long engraved maps on the ivory tusks of walrus. In the forested areas of the northeast indigenous peoples both painted and inscribed maps on the inside of birch bark. These maps were used to make voyages along the rivers and waterways. Sometimes they were left as cartographic messages along the trail, attached to blazed sticks. These were ephemeral maps, but one was preserved because it was mounted and framed soon after it was made. An inscription above the map, found by Captain Bainbridge of the Royal Engineers between Ottawa and Lake Huron in May 1841, notes that it was 'made by Indians on birch-bark and attached to a tree to shew their route to others following them'. Turning further south Barbara Mundy has identified four types of Mesoamerican maps: terrestrial maps with a historical narrative; terrestrial maps without a narrative, including property maps, city plans and itineraries; cosmographical maps and celestial maps.[2]

Mesoamerican mapmakers made maps that depicted the geography and history of the community. The maps were both celebration and record. The founding of the community and important battles were shown on the map. A sixteenth-century map now in the British Museum depicts an area around the town of Metlatoyuca. The human figures in the middle of the map, linked with a rope symbol, show the lineage of important families. The map embodies conceptions of both space and time. Cartographies were both a geographic representation and a historical narrative.

An example of a terrestrial map without a complex historical narrative is the map in the *Codex Kingsborough* of the town of Tepetlaoxtoc. The map shows trees, rivers with eddies, paths with footprints. Around the maps are toponymic hieroglyphs that are names of places and settlements. One pre-Hispanic map, contained in a Mixtec screen, the *Codex Nuttall*, shows a U-shaped valley with two rivers in the valley floor and combines topography and symbols in an elaborate design.[3]

Cosmographical maps in Mesoamerica represent the cosmos as three primary layers of earth, sky and underworld, sometimes further broken down in sublayers. The earth is divided into four quadrants. At the centre of the earth the world tree is the primary vertical axis through the three layers. World trees that hold up the world are also found in the four quadrants. These trees are shown in the map of the cosmos in the *Codex Fejérváry-Mayer*. Mesoamerican societies had an intense interest in the movement of the stars and celestial maps on stone and paper were drawn with incredible accuracy. Particular attention was paid to the movement of Venus, believed to be the god Quetzalcoatl. Celestial maps are found throughout the New World. The tribes of the western plains drew star charts on animal skin. The night sky, undimmed by modern lighting, was a backdrop to both careful measurement and imaginative storytelling.

While there are only a few examples of maps in the upland zone of South America, William Gustav Gartner argues that from around 900 BCE to CE 1532 many spatial representations emerged. Dominant field systems were either radial, emanating from a central settlement point, or parallel strips. These radial and parallel patterns formed the design template of ceramics, stone and textiles. In religious mapping ceremonies, landscape features embodied in amulets were spread out on a flat surface around the representation of a sacred place, *huaca*.[4]

The Incas created a variety of maps. One of the most fascinating is the *khipu*, a knotted string composed of a primary cord and secondary strings: it was used as a counting and as a mapping system. The different pendants on individual *khipus* represent kin groups associated with different territories.

Throughout South, Central and North America, pre-Columbian maps came in many different guises. Textiles, statues, ornaments, decorations, houses and even whole cities were laid out as maps of the terrestrial and celestial world. Maps were not only made for specific purposes such as drawing boundaries or identifying landownership but were also made *and* read as part of a cosmology that links earth and sky, here and now, geography and history.

Maps in the Columbian encounter

The Columbian encounter varies over centuries as Europeans made their way across the vast continent, sometimes quickly, at other times more slowly. In its earlier years, however, maps were made by Europeans who were influenced by indigenous spatial conceptions and incorporated them into their maps. In the Cortés map of Tenochtitlan, printed in 1526, both European and indigenous forms of mapmaking are employed in a complex juxtaposition of plan, bird's eye view and pictorial representation.

One of the earliest surviving maps of North America is the 1544 De Soto manuscript map of the Gulf of Mexico and Florida (illus. 4). The map depicts the rivers flowing into the sea. Hernando de Soto (1500–1542) arrived in what is now Tampa Bay on 27 May 1539 as head of an expedition of 600 men searching for gold. Only about half the men survived the journey through the southwest, not including de Soto, and by 1543 the remaining members of the party sailed on seven barges into a settlement in Mexico. The map was probably drawn in Seville by mapmaker Alonzo de Santa Cruz from material provided by the survivors. The map is made from memory; some of the earliest areas visited by the expedition, most easily forgotten, are not recorded. It also reflects a strong indigenous contribution. Native Americans depicted rivers, including the overland routes that connected them, as one continuous line. This indigenous convention is evident in the de Soto map where the depicted rivers show both water and land courses as one continuous river. This 'technically incorrect' hydrography depiction is incorporated into European maps of the region for another 150 years.[5]

A map of Guiana in South America was made by Sir Walter Ralegh around 1595. Searching for the mythical city of El Dorado and the prospect of a huge supply of gold, Ralegh used native informants in drawing this map, which shows El Dorado on the edge of the equally mythical Lake Manoa. There is an obvious Amerindian influence in this map: south is at the top of the page and many native settlements are noted. The lake in the map is infused with animistic conceptions of places as compared to the European notion of the environment as inert. The lake, prominently centred, looks strangely animal-like – the 'creature in the map', to cite the title of a book on Ralegh's search for El Dorado by Charles Nicholl (1995). Its consistent scale of 1 inch to 50 miles makes this map a scientific document representing both European and Amerindian spatial imaginations.[6]

Juan de Oñate, Governor of New Mexico, explored the Southern Plains and southwest in a series of journeys from 1598 to 1605. He was very reliant on indigenous people. In 1602 a captured Indian from the

Plains was questioned extensively in Mexico City. The result was a map of an area extending to over 100,000 square miles. A space–time map, it showed places in relation to travel times, containing detailed information on social and physical geography.[7]

The earliest known map to acknowledge explicitly its indigenous roots is the Velasco map of 1611. We are not sure of the mapmaker; the name of the map refers to the Spanish ambassador to England, Don Alonso de Velasco, who sent the map to the king of Spain in 1611. In an accompanying letter Velasco informs the king that it is a copy of a map made by a surveyor sent by the king of England to map the English territory in America. The map was probably made in 1610 and depicts the coast of North America from Newfoundland to Cape Fear. It is in colour and written in the northeast of the map is the phrase 'All the blue is done by the relations of the Indians'. The mapmaker has signified his reliance on Native Americans. The Velasco map is probably the basis for the general outline of John Smith's 1612 map of Virginia.

When the Europeans came to the New World, they needed spatial information in order to get around, identify peoples and locate resources. The indigenous peoples drew many maps, most of them ephemeral quick sketches in the dirt or hand gestures in the air. Some were undertaken as performances. Here are just two examples.

In December 1607 John Smith was captured by the Pamunkeys in what is now called Virginia. In his 1612 *A Map of Virginia* he describes the scene,

> First they made a fire in a house. About this fire set 7 Priests setting him [Smith] by them; and about the fire, they made circle of meal. That done the chiefe Priest . . . began to shake his rattle; and the rest followed him in his song. At the end of the song, he laid down 5 or 3 graines of wheat, and so continued counting his songs by the graines, till 3 times they incircled the fire. Then they divide the graines by certaine numbers with little stickes, laying downe at the ende of every song a little stick . . . In this manner, they sat 8, 10 or 12 houres without cease, with such strange stretching of their armes. And violent passions and gestures as might well seem strange.[8]

No reason is given for the ceremony but in a later description of the same event, in *The Generall Historie* of 1624, Smith notes,

Three dayes they used this ceremony; the meaning whereof they told him, was to know if he intended them well or no. The circle of meale signified their Country, the circles of corne the bounds of the Sea, and the stickes his Country. They imagine the world to be flat and round . . . and they in the middest.[9]

We have a visual record of the event in the visual decoration that surrounds a map in the 1624 *Generall Historie* (illus. 5 and 6). The ceremony was the enactment of a circular cosmography, a map of the world with the fire representing their village and concentric circles of the local territory (meal), the coastline (corn) and the edge of the known world (corn). Smith's country was signified by the sticks in the third circle. Some argue that the ceremony was enacted to control how the English (the sticks) entered the Powhatan's territory (the grains of corn).[10]

Several years later, in 1621, and further north, pilgrim settlers sought to form alliances with local tribes. Edward Winslow, Stephen

5 Map of Virginia from John Smith's *Generall Historie of Virginia . . .* (London, 1624).

6 Detail of Smith's 1624 map of Virginia.

Hopkins and a Native American, Squanto, left the tiny and only recently established settlement of Plymouth to travel inland. After a 40-mile trek, they reached the Pokanoket village of Sowam (close to present-day Warren, Rhode Island). To cement their alliance they gave the local chief Massosoit gifts of a copper chain and a coat. In response the chief gave a speech to the visitors and the assembled tribe in which he named all the villages that were under his control. As each village was named, the assembled Pokanoket warriors called out that he controlled this village and its inhabitants would trade with the newcomers. This performance went on for some time, as more than 30 villages were cited by the chief, each naming followed by the same refrain. Winslow noted that while it was delightful it was tedious. He was unaware, perhaps, that he had just witnessed a ritual that mapped the chief's power. The speech and the common refrain collectively bound the territorial extent of the many villages under his control. The geography of the chief's power and the spatial extent of his reach were articulated in a performance of rhetorical mapmaking.[11]

In the immediate post-conquest period maps were produced by the indigenous people for the Europeans. In order to gain control of the territory, obtain their bearings and understand the local environments,

Europeans called upon indigenous peoples, sometimes forcibly, sometimes for payment, to make maps. In 1683 the English trader Robert Livingston obtained a map of the Susquehanna River from three Indians, two Cayugas, Ackentjaecon and Kaejaegoehe, and one unnamed Susquehannock. The map, enhanced and redrawn by Livingston, depicts a section of the river that was well known to the Indians but little known to the colonists. The map depicts the course of the river as well as tribal territories, falls, portages and villages, and is very sophisticated as it scales distance by travel time. The map provided the opportunity for increased trade and contact.

Francis Nicholson, the governor of Maryland, South Carolina and Virginia from 1721 to 1728, was interested in native cartographic knowledge. Around 1721 he was presented with two indigenous maps: a Chickasaw map on deer skin depicting the Indian nations between South Carolina and the Mississippi and a Catawba map showing Indian tribes in South Carolina. The cartographers, each referred to as an unnamed 'Indian Cacique', produced highly abstract depictions of intertribal connections. The Chickasaw map covers a vast area of more than a 386,000 square miles. Indigenous maps were not just of local areas; they also reflected wider regional knowledge and accumulated layers of spatial information. The Catawba map shows the tribes in the region represented by circles and connected by trade routes. The mapmaker, a Catawba, marginalized the rival tribes, the Chickasaw and Cherokee, by placing them at the edge of the map, at the end of the network of connections and peripheral to the central representation of the British and the Catawba tribes. The grid shown on the left of the map in illustration 7 is the street pattern of Charlestown (later Charleston, South Carolina).

As Europeans pushed further into the interior of North America, they relied even more heavily on native informants and mapmakers. We have scattered evidence of what must have been extensive cartographic encounters. Many European maps were derived from Amerindian maps. In some cases the debt was noted, even in printed maps. In Philippe Buache's 1753 printed map *Carte physique des terreins les plus élevés de la partie occidentale du Canada*, the Amerindian presence is not only clear in the elongated depiction of the river systems, but the mapmaker also noted at the top of the page, 'Réduction de la Carte tracée par le Sauvage Ochagach et autres'. The map produced in

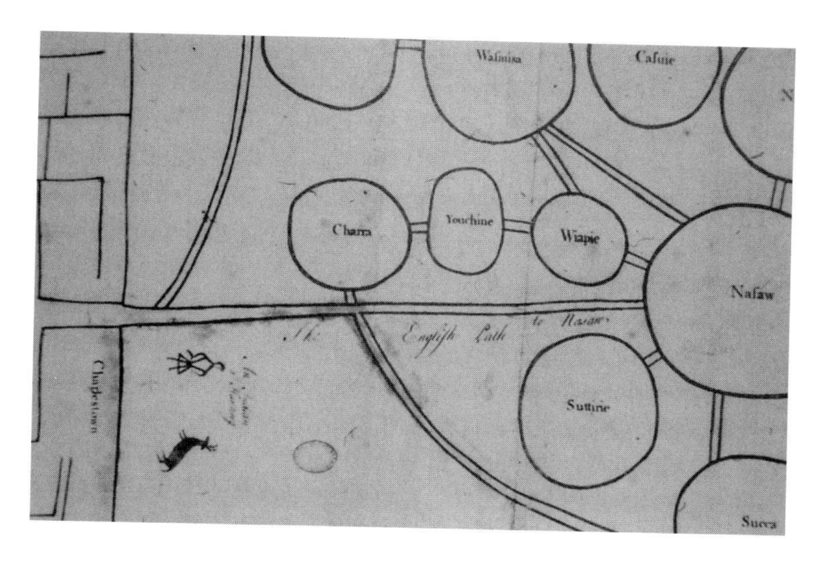

7 Detail of the 1724 Catawba map.

Paris, a centre of representation and calculation, pays homage to the indigenous sources. Thomas Kitchin's *A New Map of the Cherokee Nation* was printed in the *London Magazine* in February 1760. At the bottom of the page is the note, 'Engrav'd from an Indian draught.'

Europeans also transcribed Amerindian maps. Fur traders were at the most extreme point of contact, far from white settlements. They relied on Native Americans to provide geographical knowledge as well as pelts. Peter Pond was an American fur trader who while travelling in the Great Lakes region amassed considerable information from Native Americans. In the winter of 1784–5 he drew up this information in Montreal in a celebrated map that depicts the land from the Hudson Bay to the Rocky Mountains. A simplified copy of the map was published in the March 1790 issue of the *Gentleman's Magazine*. In 1795 a copy of the original map was presented to the Lieutenant Governor of Quebec and the US Congress with the note that the map was based on 'his own Discoveries and from the reports of the Indians'.

Peter Fidler worked for the Hudson's Bay Company as a surveyor. He was based at Chesterfield House east of present-day Alberta, at the very edge of Company territorial control and geographical knowledge. He relied heavily on local informants. In 1802 he persuaded a Blackfoot chief, Ki oo cus, to make a map of the Missouri basin. The chief highlighted the vegetation across the region and sources of food with notations such as 'plenty of berries'. Fidler also persuaded another Blackfoot chief, Ac ko mok ki, to draw maps of the region. In 1801–2 the chief drew two maps of the upper Missouri and Rocky Mountains, most likely as lines on the earth or in the snow; they were transcribed by Fidler in his journal. In the 1801 map Ac ko mok ki drew a vast area of approximately around 200,000 square miles from the Great Plains to the Pacific Ocean marking the Rocky Mountains and the path of the Columbia, Missouri and Snake Rivers as well as the sites of Indian villages. Fidler drew the map and sent it back to London noting that it showed areas 'hitherto unknown to Europeans'. Ac ko mok ki's map, transcribed by Fidler, is one of the great maps of the American West (illus. 8). It depicts a vast physical geography and a subtle human geography. Circles represent native settlements with numbers indicating the number of tents (illus. 9). The map reveals a complex hydrology and exhibits a keen knowledge of topography and land use – the line between forest and grassland is marked – as well as a detailed understanding of the human geography of a settled land. Fidler sent his maps back to London, where they were incorporated into Aaron Arrowsmith's 1802

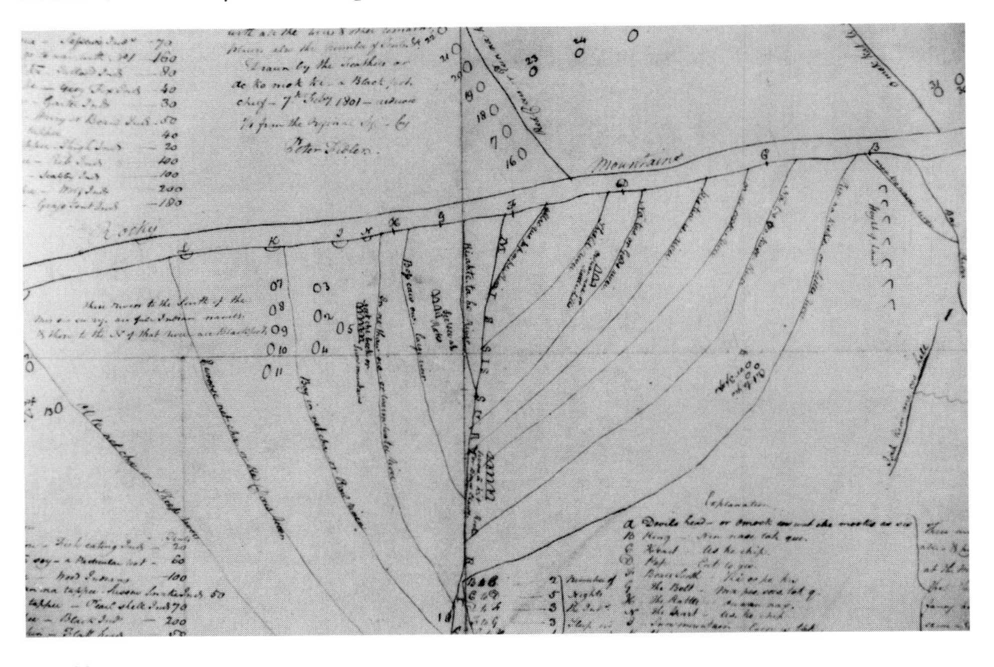

8 Ac ko mok ki's 'Map of The West', 1801.

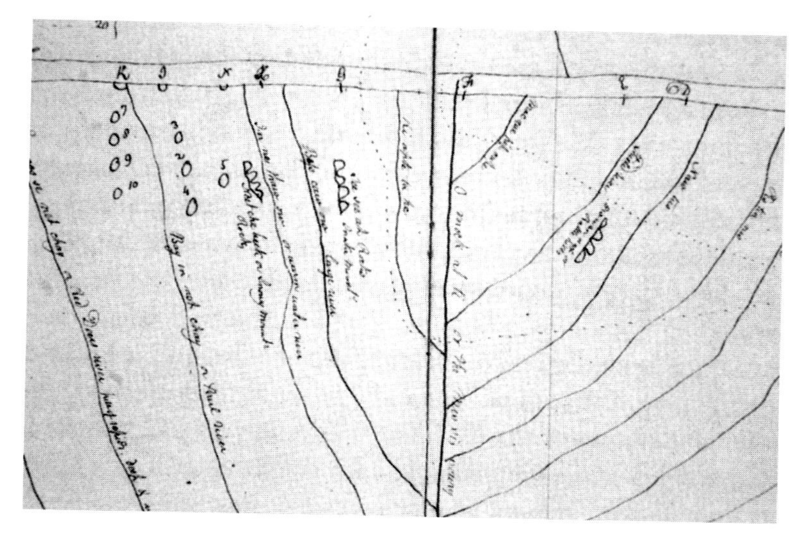

9 Detail from a later revised copy of Ac ko mok ki's 'Map of The West'.

edition of his printed map of North America, *A Map Exhibiting All the New Discoveries in the Interior Parts of North America*, showing the course of the Missouri River from the Rocky Mountains to St Louis. It indicated to Thomas Jefferson that a route was possible from St Louis up the Missouri through the Rockies to the Pacific. Lewis and Clark used the Arrowsmith map as they made their way to the Pacific.

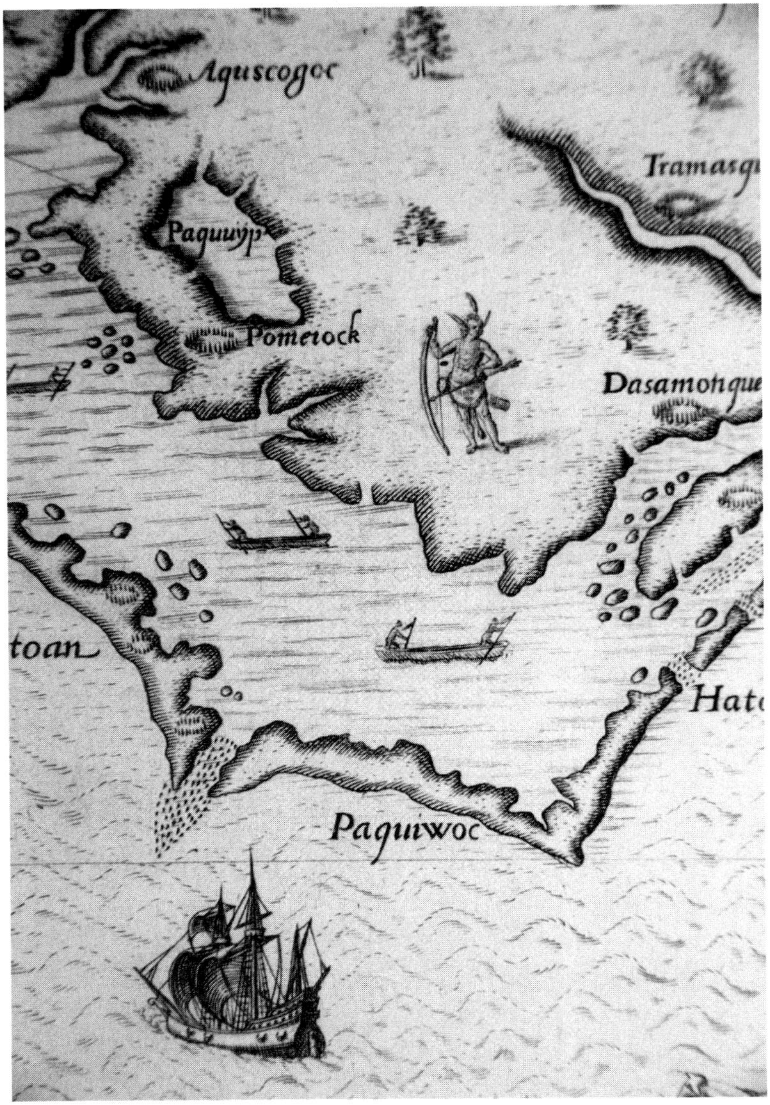

Aguscogoc
Paquuyp
Pomeiock
Paquiwoc

PART II | colonial cartographies

[O]n the ninth of April, with all possible solemnity, we performed the ceremony of planting the cross and raising the arms of France. After we chanted the hymn of the church, 'Vexilla Regis' and the Te Deum, the Sieur de la Salle in the name of his majesty took possession of that river, of all the rivers that enter it and of all the country watered by them. An authentic act was drawn up, signed by all of us there, and, amid volley from all our muskets, a leaden plate subscribed with the arms of France and the names of those who had just made the discovery was deposited in the earth.

Father Zenobius Membre, 1682

10 A detail from John White's map of Virginia, published in Theodor de Bry's *Admiranda narratio fida tamen, de commodis et incolarum ritibus Virginiae* (Frankfurt, 1590).

Fort van Nassoueen is binnen de wallen 58 voeten wijt int vierkant / de gracht is wydt 18 voeten

NAV

E

SE

Riviere vanden Vorst Mauritius

MAHICANS

MAKIMANES

R vanden roden bergh

Versche rivier

Rodnijns

Archipelago

Hellegatt

De Keer

SANGICANS

Sandbay

Sandpunt

ACHVKES

3 | Encounters in a Settled Land

Cartographic encounters were collaborations between newcomers and indigenous people. Each side had something to gain and something to lose. The Europeans needed geographical knowledge. They lacked the spatial information necessary to move around in and to map this new land. The indigenous peoples had sophisticated spatial knowledge and cartographic abilities.

In the encounter the colonists had much to offer. They had goods such as the beads and mirrors that the indigenous people could trade. They had powerful technologies such as axes, blankets, fish-hooks, guns and other metal goods that made life easier. Armed with European tools, hunters no longer needed to track game with bows and arrows, they could kill them with guns and metal traps. The newcomers provided commodities that made life easier and trade more lucrative. The new technologies and new trading opportunities overturned traditional Indian–environment relations. In the early colonial period the whites wanted fur and information and in return they gave goods, guns and alcohol. Calvin Martin describes how the Micmac and Ojibwa tribes, even as early as the middle of the seventeenth century, were principally responsible for the over-hunting of fur-bearing animals in eastern Canada. The new, more efficient, hunting technology, the insatiable white demand and the lack of moral sanctions in the wake of the collapse of traditional belief systems and the spiritual turmoil

11 Detail from Adriaen Block's 1614 map.

associated with contact with the Europeans, all led to the unrestrained slaughter of game. The traditional keepers of the game became the new annihilators.[1]

The Europeans had power, their very presence a remarkable achievement to the people of the New World. 'I had heard marvelous things of this people', noted Charles Alexander Eastman, a Santee Sioux who never saw a white until he was sixteen years of age. 'In some things we despised them; in others we regarded them as *wakan* (mysterious) a race whose power bordered upon the supernatural.'[2]

For the Europeans the indigenous people represented suppliers of food and trading partners as well as a source of vital geographical information. However, to trade with the indigenous peoples was to give them access to the new technology. The guns that were traded could be turned against the colonists. And to strike an alliance with one group could invoke the enmity of another. The complex geopolitics of the New World was a minefield that had to be negotiated with care.

In the beginning the newcomers were novelties, scarcely threatening. A folktale from the Lipan Apache, as told by Percy Bigmouth, tells of when the tribe first saw whites. A tribal council debated whether to kill them or let them live. 'Let's not kill them', someone said, 'they may be a help to us some day.'

> It was spring. The Lipan gave them some pumpkin seeds and seed corn and told them how to use it. The people took it and after that they got along all right. They raised a little corn and some pumpkins. They started a new life. Later on the Lipan left for a while. When they returned, the white people were getting along very well. The Lipan gave them venison. They were getting along very well. After that, they began to get thick.[3]

Europeans offered access to expanding trading opportunities, advanced technologies and the possibility of alliances in a fluid and ever-changing geopolitics. As news of their presence swept thought the continent, Native American leaders realized that the Europeans represented a danger but also an opportunity to ally themselves with a superior force in their continual struggles with their traditional enemies and rivals. Collaboration with the Europeans was an opportunity to vanquish traditional foes. The Europeans had the firepower to wage a winning war.

As their numbers 'thickened' the whites took over the land and no longer needed the indigenous people. As Europeans gained more information and mapped the land the indigenous people became a hindrance, a block to settlement. The cartographic encounter became a holocaust. As the frontier moved further west and the whites spread the indigenous collaborators, their help once so vital, now became irrelevant. Once the information from the local peoples was obtained and processed into popular knowledge and scientific understanding, the peoples were no longer needed. Annihilation replaced collaboration behind the moving frontier. The more information and help the Europeans obtained from the indigenous people, the less they needed them. As they moved west, Europeans became Americans, and the collaboration became even more unequal. With more land than sea behind them, the new Americans were now extending their control rather than establishing a foothold. For the indigenous peoples the seeds of destruction were sown in the collaboration. But that is the end of our tale; let us return to the beginning.

The New World was not an uninhabited wilderness, it was populated by a rich variety of people. Estimates of the population of the New World on the eve of Columbus's landfall vary from 50 million to 120 million. Recovering accurate numbers is made difficult by the lack of records, the passage of time and the politics involved. Earlier studies tend to undercount the numbers of indigenous people and more recent studies tend to go for higher estimates.[4] What is not at issue is that the continent was filled with people. From the coast to the desert interior, from the humid plains to the icy mountains, the New World was a populated continent comprising some hunter-gatherers, some farmers, some living in small villages, others in large cities. The New World in 1491 was inhabited, occupied. The early colonial maps provide witness to this humanized landscape.

The importance of Native Americans to the European understanding of the New World is evidenced by their depiction in early European maps. The Native American presence was recorded as an important geopolitical fact, and the maps document this in rich detail. Early maps took great pains to record cartographically tribal variations because it was important to be aware of geopolitics. The recording of

Native Americans in early maps speaks to both the settled nature of the land and the crucial need to document subtleties of tribal geopolitics. The maps also embody cartographic encounters. Let us consider two examples, the so-called Block and Delisle maps.

The Block Map

One of the earliest European maps, which depicts what is now New England and New York, is a Dutch map, dated 1614, entitled *Map of New Netherlands*. It first appears in the historical record at a meeting of the States General of the United Provinces in The Hague on 11 October 1614.

A little historical background: in the sixteenth and seventeenth centuries there was intense rivalry between the European powers for overseas trade and territory. The big prize for the Dutch merchants was Asia and the East Indies, which contained spices, timber, jewels and other precious commodities that enabled merchants to make big profits. Trading companies were formed to pool resources and share risks. In 1602 the Dutch East India Company was formed, becoming hugely successful, paying annual dividends of never less than 20 per cent and often 50 per cent.

In order to find a *western* route to the East Indies, the Company signed a contract with the English navigator Henry Hudson in 1609. They furnished him with a vessel, the *Half Moon*, which set sail on 25 March 1609. Progress was slow but eventually the ship reached the Maine coast. Hudson sailed south looking for the passage to the Indies. Unsuccessful, he returned northward, anchoring in Chesapeake Bay. He continued north and on 2 September in the same year, the ship cast anchor off Sandy Hook. The next morning members of the crew went out in boats and made contact with the local people.

Robert Juet, an officer on the *Half Moon*, wrote a journal, which was subsequently printed. Juet notes the contact between the ship and the indigenous inhabitants. At a number of points he acknowledges the sharing of local knowledge: 'So at three of the clocke in the afternoone they [Native-Americans] came aboard, and brought Tobacco, and more Beades, and gave them to our Master, and made an Oration, and shewed him all the Countrey round about.'[5] The phrase 'shewed him all the Countrey' was used often by Juet and indicates the exchange of information between the locals and the explorers. Four days later Juet notes: 'This morning two Canoes came up the River where we first found loving people, and in one of them was an old man . . . He brought another old man with him, which brought more ropes of Beades, and gave them to our Master, and shewed him all the Countrey.'[6]

The *Half Moon* spent ten days in the area before sailing up the river that we now call the Hudson and reached as far as present-day Albany before the water became too shallow for further progress. Hudson sailed back down the river trading with the local people, making surveys and soundings. On 4 October the *Half Moon* set sail for home. On the return journey Hudson's ship put into Dartmouth in southwest England. Hudson was seized by the authorities and charged with entering foreign service without permission of the English king. His maps and reports were confiscated. Hudson did not make it back to the Dutch Republic, but the information about the discovery certainly did. Here was a region of North America where no other European power had a presence and where there was a plentiful supply of beaver, an extremely valuable commodity at the time.

The map that the States General pondered at their October meeting in 1614 is now called the Block Map (illus. 11, 12), after Captain Adriaen Block, who presented the map to the merchants. It is based on Block's four voyages to North America in the years 1611 to 1614. The Block Map is a map of the entire coast from Maine to Virginia at a scale of one-and-a-half inches to one degree of latitude. The map is pen and ink and watercolour on vellum.[7] It is the first map to contain the words 'New Netherlands' and also the first to show Manhattan as a separate island. Most of the information in the Block Map was gleaned from native informants, who gave names of tribes and physical configurations. Most of the inland detail is around the mouth of the Hudson. Tribal names are noted and there is a crude depiction of forests and settlements. Across the map the names of local tribes are depicted: Mahicans, Makimanes, Nahicans, Morhicans, Sangicans and Manhates. In a detail of the map (illus. 11), Manhattan is depicted as the small island of Manhates. In another part of the map (illus. 12), one tribe, the Mohawk, listed as Maquaas on the map, is noted as comprising canoe-makers. The Block Map does not show an empty wilderness waiting for the European settler, but a human

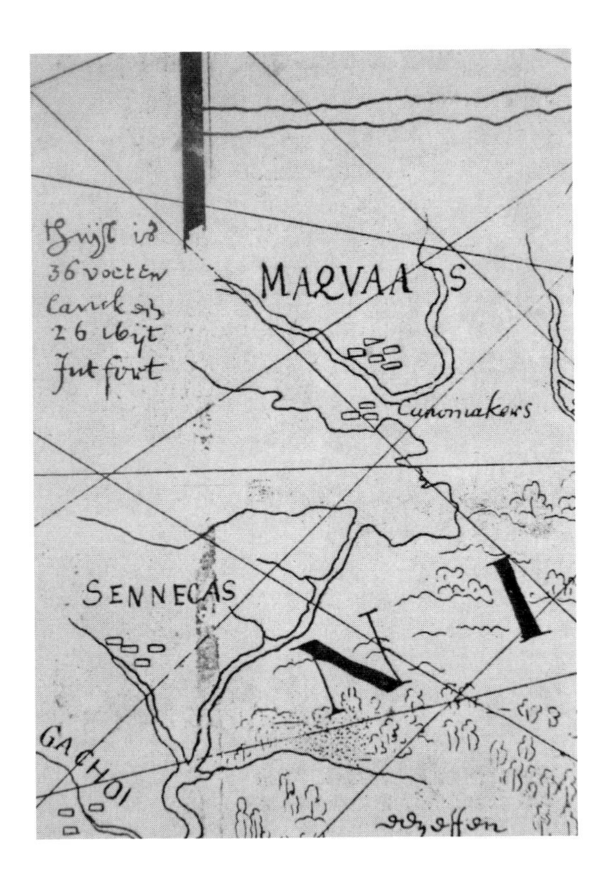

12 Detail from Adriaen Block's 1614 map.

landscape with rich evidence of settlement: houses are marked as well as the dense patchwork of varied tribal names and affiliations. The map, for all its crude outlines and hazy geography, is a reminder that people lived here. A human presence is noted in some detail. The information was obtained from communication with the indigenous inhabitants, for whom human settlements were an important part of their geographical knowledge. It was also important for the Dutch to know who lived where. The map depicts a populated landscape because the Dutch were looking to trade, and the map provided a spatial ordering of potential trading partners. So it was not an innocent ordering. Across the map, inscribed in block capital letters, is 'New Netherlands', an act of branding that is also a gesture of appropriation.

The Delisle Map

The colonial maps are not so much simple descriptions of territory as acts of domination, claims to ownership in a cartographic legitimation of control. The mapping of the New World was never innocent of political agendas. The New World was appropriated as well as understood through mapping. Maps became the documents of both scientific understanding and political control. To map was to incorporate into everyday knowledge, scientific understanding and political ambition. The colonial maps that enabled mobility were also acts of military surveillance, claims to landownership and representations of the native other. They also bear witness to a major native contribution. The European cartographic depiction of the land was the product of a series of cartographic encounters between two peoples: the indigenous people with detailed spatial knowledge of the land and the colonialists seeking to obtain this land.

One of the most important early eighteenth-century maps of North America is Guillaume Delisle's *Carte de la Lousiane et du Cours du Mississipi*. Made in Paris in 1718 by the official geographer to the king, the map depicts the vast interior of North America from the Great Lakes to the Gulf of Mexico and from the Atlantic coast to the Rockies.[8] It draws upon the maps, journeys and writings of many French explorers, including Jacques Marquette, Louis Joliet, René-Robert Cavelier, Sieur de La Salle, and Louis Hennepin. The French explorers were very dependent upon Native Americans for supplies, information and guiding through the unfamiliar territory. The French moved from their base in New France, around the St Lawrence, southwards via the rivers to the mouth of the Mississippi in search of allies, furs, control over the giant river basin and easier access to the sea. In 1673 Louis Joliet, along with Father Marquette and others, travelled south as far as the Arkansas River. Several years later, in April 1682, La Salle planted a cross at the mouth of the Mississippi to mark French possession of the entire river basin.[9]

We have a record of the earlier part of La Salle's trip. The priest Louis Hennepin accompanied the expedition and wrote it up as *Description of Louisiana*, first published in Paris in 1683. It describes many examples of the French reliance on Native Americans. These include the direct transfer of geographical information, as in this case from a young Illinois man:

> This young man traced for us with coal, a pretty exact map, assuring us that he had been everywhere in his periagua [canoe]; that there was not down to the sea, which the Indians call the great lake, either falls or rapids. But that as this river became very broad, there were in some places sand banks and mud which barred a part of it. He also told us the name of the nations that lived on its bank, and of the rivers which it receives. I wrote them down.[10]

Later in the journey,

> Toward the end of September, having no implements to begin an establishment, we resolved to tell these people that for their benefit, we would have to return to the French settlement. The great chief of the Issati or, Nadouessiouz consented, and traced in pencil on paper I gave him, the route we were to take for four hundred leagues of the way. With this chart, we set out, eight Frenchmen, in two canoes and descended the rivers of St Francis and Colbert.[11]

It is interesting to note, as this excerpt reveals, that the cartographic encounters were made easier by the writing tools carried by the Europeans. The pens and paper provided an important method of communication.

The French traded with the locals for food and the information necessary to negotiate the complex portage systems that linked river systems. In some cases they followed the Native Americans' lead. In 1679, at a place close to present-day South Bend, Indiana, Hennepin notes,

> We rejoined our party the next day at the portage where Father Gabriel had made several crosses on the trees, that we might recognize it. We found there a number of buffalo horns and the carcasses of those animals, and some canoes that the Indians had made, of buffalo skins to cross the river with their load of meat.[12]

In other places they relied directly on Native Americans. By the winter of 1679 La Salle's party was running out of food and direction. Around present-day Peoria, Illinois, they came across an empty Illinois village; its people were off on a winter hunting trip. They stole twenty bushels of corn from the village, knowing full well they were stealing a vital resource. Hennepin notes 'This stock is extremely precious in their eyes, and you could not give them greater offense than by touching it in their absence. Nevertheless there was no possibility of our risking a further descent without food.'[13] Several days later they came across the tribe and La Salle, according to Hennepin, realizing the delicacy of the situation, told them that if they wanted their corn back either he would return it and then trade his axes for corn with another tribe, the Osages, or they could make the trade after the fact; La Salle's party would keep the corn in return for axes he would give them. He sweetened the deal by promising to protect them from their enemies and provide a steady supply of the much-prized trade goods. However, getting these goods was difficult, and that is why he needed geographical information: easier access to the sea would make the goods more available.

> But as this enterprise required a great outlay, we wished to learn whether their river was navigable to the sea, and whether other Europeans dwelt near its mouth. The Illinois replied that they accepted all our proposals, and that they would assist us as far as they could, then they gave a description of the river Colbert or Meschasipi [Mississippi]; they told us wonders of its width, and beauty, and they assured us that the navigation was free and easy, and that there were no Europeans near its mouth.[14]

The Illinois traded corn and information in return for axes, protection from enemies and access to future trade. In this cartographic encounter the Illinois hoped to benefit materially and to position themselves better geopolitically against their traditional enemies the Osages and the fierce Iroquois. The French, in return, received much-needed food that enabled them to survive the winter and much-needed spatial information that allowed them to plan their travels. By such cartographic encounters, the Europeans made their way and mapped their course through the New World.

La Salle went on to journey through the region. In July 1684 he sailed from France with an expedition of four ships. By December three of the ships sighted the mouth of the Mississippi. The expedition turned into a fiasco. Shipwrecked further west on the Gulf of Mexico, the survivors tried to make their way back to the Mississippi. In July 1686, according to Henri Joutel, they had to rely on local informants: 'The 9th and 10th were spent in visits, and we were informed by one of the Indians that we were not far from a great river, which he describes with stick on the sand, and showed it had two branches.'[15]

13 Detail from Guillaume Delisle's map of 1718, *Carte de la Louisiane et du Cours du Mississipi.*

Delisle drew upon these and other French and Native American cartographic encounters to produce one of the most accurate maps of the river system to date. It was immensely influential and was used as a template for almost fifty years. Jefferson had a copy of the map, and it was an important source of information for the Lewis and Clark expedition. The map contains an extensive listing of Indian villages and names (see illus. 13).

Early European maps such as the Block map and the Delisle maps show a Native American presence. They were a geographical fact and a political reality. However, as the Europeans, and later the Americans, gained geographical knowledge and territorial control through these cartographic encounters, the Native Americans were moved, defeated, annihilated. They were no longer presented in maps. I have demonstrated this gradual disappearance in my earlier work (see Bibliography). The earliest maps of New York State, as in the Block map, show a strong Native American presence, but by the time the first atlas of the state was published in 1830, Native Americans are completely absent. Vital in the earlier geographical understanding of the territory, by the 1830 atlas they are unrecorded. The atlas shows a state covered in a grid of latitude and longitude, divided into civil divisions, connected by roads and canals. The land is surveyed, mapped and statistically described. And nowhere are Native Americans mentioned.[16]

Further west, the Native American cartographic presence lasted longer. Henry Schenck Tanner's large wall map, *United States of America*, published in 1829, at the detailed scale of 25 miles to the inch, represents two Americas. East of the Mississippi the map is covered with towns, canals and all the other trappings of modern society. West of the Mississippi is vacant space with a Native American presence made visible. Over time the tribes disappear, and the land is portrayed as a uniform space devoid of a Native American presence. By the end

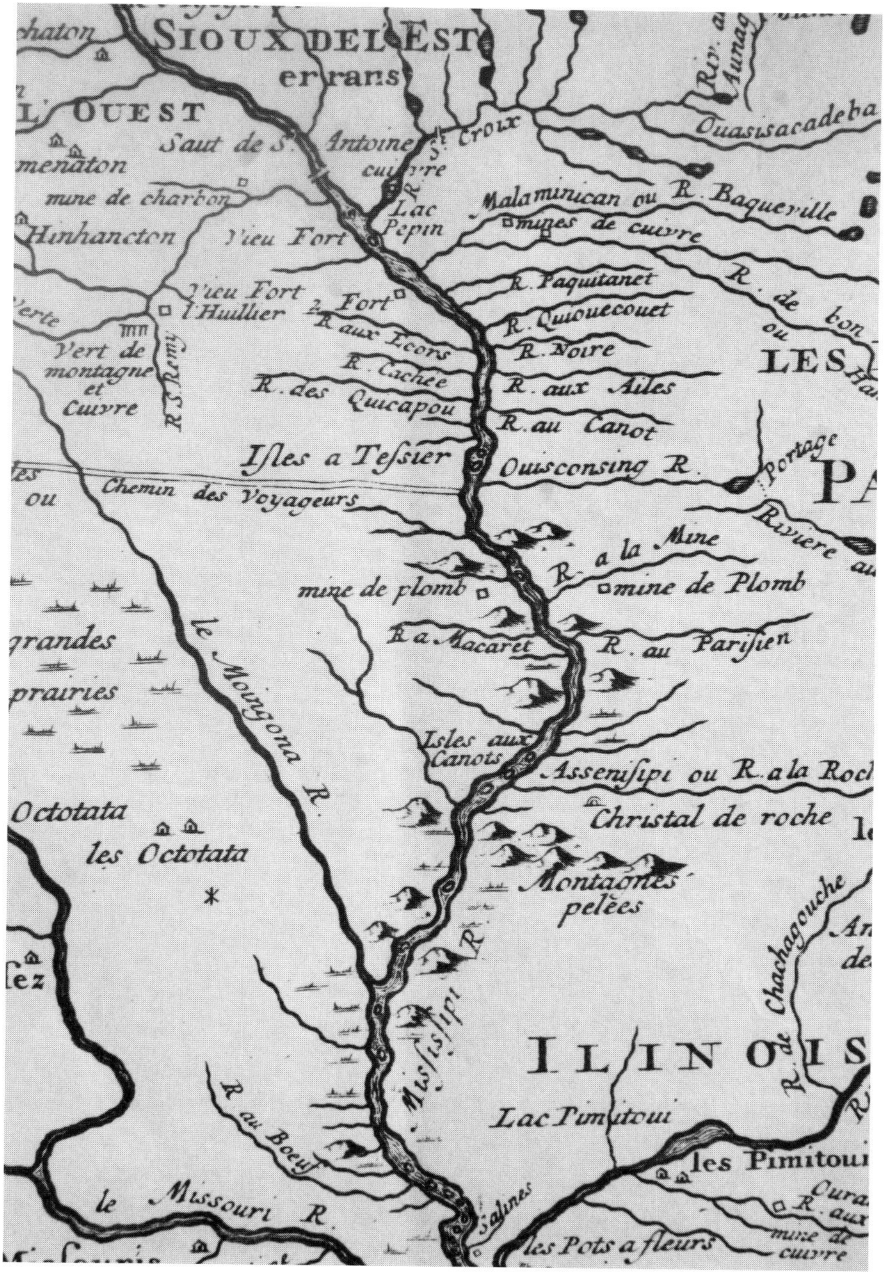

chaton
SIOUX DE L'EST
errans
L'OUEST
Saut de St. Antoine
R. St. Croix
cuivre
menaton
mine de charbon
Lac Pepin
Malaminican ou R. Baqueville
mines de cuivre
Hinhancton
Vieu Fort
Ouasisacadeba
Riv. aux Aunag
R. Paquitanet
Vieu Fort l'Huillier
2. Fort
R. Quiouecouet
R. de bon
ou
LES
R. aux Ecors
R. Noire
Vert de montagne et Cuivre
R. S. Remy
R. Cachee
R. des Quicapou
R. aux Ailes
R. au Canot
Isles a Tessier
Ouisconsing R.
Portage
PA
ou
Chemin des Voyageurs
Riviere au
R. a la Mine
mine de plomb
mine de Plomb
grandes
prairies
R. a Macaret
R. au Parisien
le Moungona R.
Isles aux Canots
Assenisipi ou R. a la Roch
Octotata
les Octotata
Christal de roche
Montagnes pelées
R. de Chachagouche
Mississipi R.
ILINOIS
ez
R. au Boeuf
Lac Pimitoui
les Pimitoui
le Missouri R.
Salines
les Pots a fleurs
R. aux
mine de cuivre

of the nineteenth century the national maps portray a unified country, the Native American presence erased and forgotten.

The cartographic encounters resulted in Native Americans losing control over their territory. No longer needed, they were moved and marginalized. Their cartographic erasure from the maps was a portent of their ultimate eradication from the land itself. Just as they were removed from the land, so they were deleted from the cartographic record. Their names sometimes live on as toponymic legacies. Manhattan remains the name of the island between the Hudson and the East River, and at least 25 of the 50 states of the union trace their names to Native American words, ideas or places. But the Native Americans that were the source of these names have long since disappeared. And to add insult to injury, the role of Native Americans in the exploration/mapping of North America is too often ignored or forgotten.

Despite the traditional view that Europeans created maps of the continent on their own, Native Americans were involved in the mapping of North America. It is more accurate to consider the notion of cartographic encounters involving Europeans and Native Americans, rather than a simple cartographic appropriation by Europeans. The mapping of the continent was underpinned by native knowledge and there is a hidden stratum of Native American geographical knowledge that is only now being uncovered.[17]

4 | Landing in a Strange Land

All the existing European accounts we have of the early arrivals to the New World tell the same story. The Europeans landed in a populated place. In order to find their way and move around the unfamiliar landscape, they relied upon the indigenous people to provide information, advice and guidance. In return the indigenous people obtained trade goods and the possibility of new alliances with powerful new forces against their long-standing enemies. This exchange was invaluable to the Europeans over both the short and long term. For the Native Americans the exchange over the short and medium term would prove useful, but over the long term it sealed their fate.

The encounters are particularly vivid for the earliest Europeans landing in, what was for them, a strange land. Consider the case of John Smith, who was part of the earliest permanent English settlement in North America, Jamestown in Virginia. The scheme was financed by the Virginia Company of London, which was looking to the New World for gold and other commodities and possibly a route to the Indies. Three ships, the *Susan Constant*, *Godspeed* and *Discovery*, with over 100 people, sailed from London on 19 December 1606. They crossed the Atlantic to the West Indies and arrived in Virginia on 26 April 1607. It was not a happy crossing. John Smith was accused of plotting against the social leaders of the expedition. Smith was an adventurer. He fought for the French against Spain, he battled for the

Dutch and joined the army of Austria fighting against the Turks. Although from a modest family, he was a well-travelled and seasoned campaigner not easily intimidated by his social betters on the voyage. Arguments erupted; he was arrested and arrived in the New World as a prisoner.

The ships dropped anchor in the Chesapeake Bay, and a party of 30 went ashore. Later that night, as they returned to the ships, they were attacked. Two of the Englishmen were badly injured. The New World of such innocent promise in London was a more dangerous place in Virginia.

The English arrived in a densely populated region. Many tribes occupied the land along the indented coastline of the Bay. The Appomattoc, Arrohattoc, Kecoughtan, Kiskiack, Mattaponi, Nansemond, Pamunkey, Paspahegh, Rappahannock, Weyanock and Younghtanund were all independent tribes. The most powerful tribe was the Powhatan, whose military empire included the territory of 30 vassal tribes including the Accomac, Chickahominy and Patawomeck. Inland the Monacans and the Mannahoacs were a powerful counterweight force to the Powhatans. The English arrived to a humanized landscape, not a wilderness, to a complex set of existing political relations, not a vacant space.

The fledgling English colony was crucially dependent on the corn, knowledge and goodwill of the tribes. The early days were spent in gift-giving, feasts and orations. The colonists sought to establish good relations. Beads, hatchets and other goods were exchanged for vital supplies. The colonists were vulnerable. Knowing little about the land they existed through the presence of Native Americans, not despite them. Within months, half the colonists were dead. Starvation, disease and exhaustion exacted a heavy toll. There was internal dissension and eventually John Smith assumed a powerful position. Smith realized that the way to survive was to gain as much information as possible from the locals.

Smith had to negotiate from a position of weakness against powerful forces that could, and sometimes did, conspire to wipe out the colonists. His earliest experiences are recorded in *A True Relation of Such Occurrences and Accidents of Note as Hath Hapned in Virginia*. This text was originally a long letter he wrote about the first winter in Jamestown. It was rushed into print in 1608, without Smith's knowledge, containing errors and some crude editing. However, it retains the immediacy of a letter and provides a more journalistic account of the experiences, whereas his later writing seems more angled for posterity than for accuracy. Despite the problems with the text, I feel it has a more direct relationship to events, especially in the dealings with Native Americans, than *The Generall Historie of Virginia* that was published in 1624.

In *A True Relation*, Smith's narrative is full of the importance, necessity and complications of dealing with the indigenous people. He depicts a whole series of cartographic encounters. So many, in fact, that I will simply list a few. On 14 May 1607 the expedition landed on the banks of the James River at a site that they named Jamestown. Just over a week later Smith notes,

> Captain Newport and my selfe with divers others, to the number of twenty two persons, set forward to discover the River, some fiftie or sixtie miles, finding it in some places broader, and in some narrower, the Countrie (for the moste part) on each side plaine high ground, with many fresh Springes, the people in all places kindley intreating us, daunsing and feasting us with strawberries, Mulberries, bread, Fish . . . for which Captain Newport kindely requited their least favours with bels, Pinnes, Needles, beades, or Glasses which so contented them that this liberallitie made them follow us from place to place, and ever kindely to respect us. In the midway staying to refresh ourselves in a little isle, four or five savages came unto us which described unto us the course of the River, and after in our journey, they often met us, trading with us for such provison as wee had, and arriving at Arsatecke, hee who, we supposed to bee the chiefe King of all the rest, moste kindely entertained us, giving us in a guide to go with us up the River to Powhatan.[1]

Within a mile their journey was blocked by a rocky part of the river:

> That night we returned to Powhatan: the next day . . . we returned to the fals, leaving a mariner in pawn with the Indians for a guide of theirs, hee that they honoured for King followed us by the river. That afternoone we trifled in looking upon the

> Rockes and river (further he would not goe) so there we erected a crosse.[2]

Reading Smith's account makes it clear that without their native informants and native guides the Englishmen were stumbling around lost and blind. They could only move around the country with the help and advice of local people. In November 1607 Smith set out to explore more of the country and to search for the continental river passage. Smith's party sailed 40 miles up the Chickahominy River. Blocked by a tree, Smith obtained a smaller canoe, 'Having 2 Indians for my guide and 2 of our own company, I set forward.'[3] Smith was separated from his own company and after a brief fight was captured. Even the meeting with Opechancanough, chief of the tribe and brother of Powhatan provided an opportunity to gather more spatial information: 'What he knew of the dominions he spared not to acquaint me with.'[4] Later, Smith was taken to meet the paramount chief of the Powhatan, variously known as Chief Powhatan, Wahunsenacawh or Wahunsunacock,

> After good deliberation he began to describe mee the Countreys beyonde the Falles, with many of the rest; confirming what not only Opechancanoyes, and an Indian which had beene prisoner to Pewhatan had tolde mee: but some called it five dayes, some sixe, some eight, where the sayde water dashed amongst many stones and rockes.[5]

For Smith, capture by the Indians provided an enforced cartographic encounter that gave him an opportunity to glean more information. And after his release, now armed with even more information of the lay of the land and the disposition of the Powhatan's allies and enemies, he was in a much better position to move around the country and secure the future of the struggling colony.

Smith turned his geographic information into a map that was published in Oxford in 1612 to accompany a description of the region, *A Map of Virginia with a Description of the Country, the Commodities, People, Government and Religion* (illus. 14). It is one of the earliest European maps of the mid-Atlantic regions. Smith orientated the map with west at the top, reflecting a mariner's entry into the region:

14 John Smith's *Map of Virginia* (Oxford, 1612).

it is surprisingly accurate and when set aside a contemporary satellite image is eerily similar.[6] It draws heavily upon Native American informants, guides and geographic descriptions and carefully delineates the coastline and the river systems as well as forests and hills. The names of the many and various tribes are inscribed on the map with their principal settlements. In the accompanying text Smith notes, 'In which Mappe observe this, that as far as you see the little Crosses on rivers, mountains, or other places, have been discovered; the rest was had by information of the Savages, and are set down according to their instructions.'[7] The crosses are clearly visible (illus. 15). Even the area within the crosses is much more the result of cartographic collaboration than Smith suggests in the map.

Smith's 1612 map, so often portrayed and celebrated as a European achievement, is, in fact, the result of cartographic encounters with the indigenous population achieved through travel and trade,

15 Detail of John Smith's 1612 *Map of Virginia.*

imprisonment and dialogue. Although Smith drew the map, the spatial knowledge on which it is based draws upon results from these encounters.

When the cartographic encounters are represented in later forms, a subtle (and sometimes not so subtle) claiming of territory is apparent along with a downgrading of the Native American role. In the 1612 map the Native American presence is made visible and obvious. Not only are the tribal names depicted prominently on the map, there are also visual references. In the top right of the map a male figure stands proudly beside his longbow. And in the top left Chief Powhatan is shown with his subjects. In 1624 Smith published *The Generall Historie of Virginia, England and The Summer Isles.* The frontispiece in the 1634 edition is a map with English/British Royalty stamped across the land and a royal inscription (illus. 16). The earlier 1612 map is now described in the text as 'a Mappe of the Countrey of Virginia now planted'. The period from 1612 to 1632 marks an erasure of the Native Americans as the colonial space is now inscribed and claimed as well as described.

THE
GENERALL HISTORIE
OF
Virginia, New-England, and the Summer
Isles: with the names of the Adventurers,
Planters, and Governours from their
first beginning. An: 1584. to this
present 1624.
WITH THE PROCEEDINGS OF THOSE SEVERALL COLONIES
and the Accidents that befell them in all their
Journyes and Discoveries.
Also the Maps and Descriptions of all those
Countryes, their Commodities, people,
Government, Customes, and Religion
yet knowne.
DIVIDED INTO SIXE BOOKES.
By Captaine IOHN SMITH sometymes Governour
in those Countryes & Admirall
of New England.
LONDON.
Printed by I.D. and
I.H. for Michael
Sparkes.
1624.

16 Frontispiece of the 1634 edition of John Smith's *Generall Historie of Virginia.*

Jamestown became a base for subsequent English explorers. Here is a typical encounter. In 1634 Captain Thomas Yong and his two ships, one of them leaking badly, arrive at the settlement. He had sailed from London on a royal commission to look for the Northwest passage. The ship is repaired and on 20 July Yong sails further north into what is now Delaware and New Jersey. Four days later, 'When I gott into the Bay, I came ashore, on the Southwest part of the Bay, to see if I could speake with any of the Natives, and to learne what I could of them concerning this Bay, and the course thereof.'[8] On the 27th he persuaded a 'Native' to come aboard his ship:

> I entertained him courteously, and gave him biscuit to eat and strong water to drink . . . I also gave him some trifles, such as knives and beades and a hatchet of which he was wonderfully glad. Then I began to enquire of him (by my Interpreter who understood the language) how farr the sea ran, who answered me that not farr above that place I should meet with fresh water, and that the River ranne up very farre into the land . . . He told me further that the people of that River were at warre with a certaine Nation called the Minquaos, who had killed many of them, destroyed their corne and burned their houses; insomuch as that the Inhabitants had wholly left that side of the River, which was next to their enemies and had retired themselves on the other side farre up into the woods.[9]

Yong receives information on the basic physical geography as well as a lesson in local political history and current geopolitics. A couple of days later Yong comes across another tribe:

> I enquired of this king how farre this River ranne up into the Countrey, and whither it werr navigable or no, he told me it ranne a great way up, and that I might gow with my shippe, till I came to a certaine place, where the rockes ranne cleane across the River [now Trenton, New Jersey] . . . I then desired him to lend me a pilott to goe up to that place, which he most willingly granted.[10]

Through these cartographic encounters, the English explorers received vital information while indigenous groups obtained trade goods and geopolitical information (the explorers told them of other regions and tribes), and the possibility of alliances with the English against their enemies. In the short to medium term the encounters could be and often were profitable to the Native Americans. But in the long term they gave up much more valuable spatial information to the peoples who mapped and appropriated the land. In 1607 more than 30 Algonkian-speaking tribes of more than 15,000 people occupied much of coastal Virginia while the Monacans and Mannahoacs lived in the west of the state. Four-hundred years later, in 2007, the state of Virginia recognizes only eight tribes that together account for just 4,475 people. Only 1,359 acres are in Indian reservations. The once mighty Powhatan became extinct in the eighteenth century and their language is lost forever.

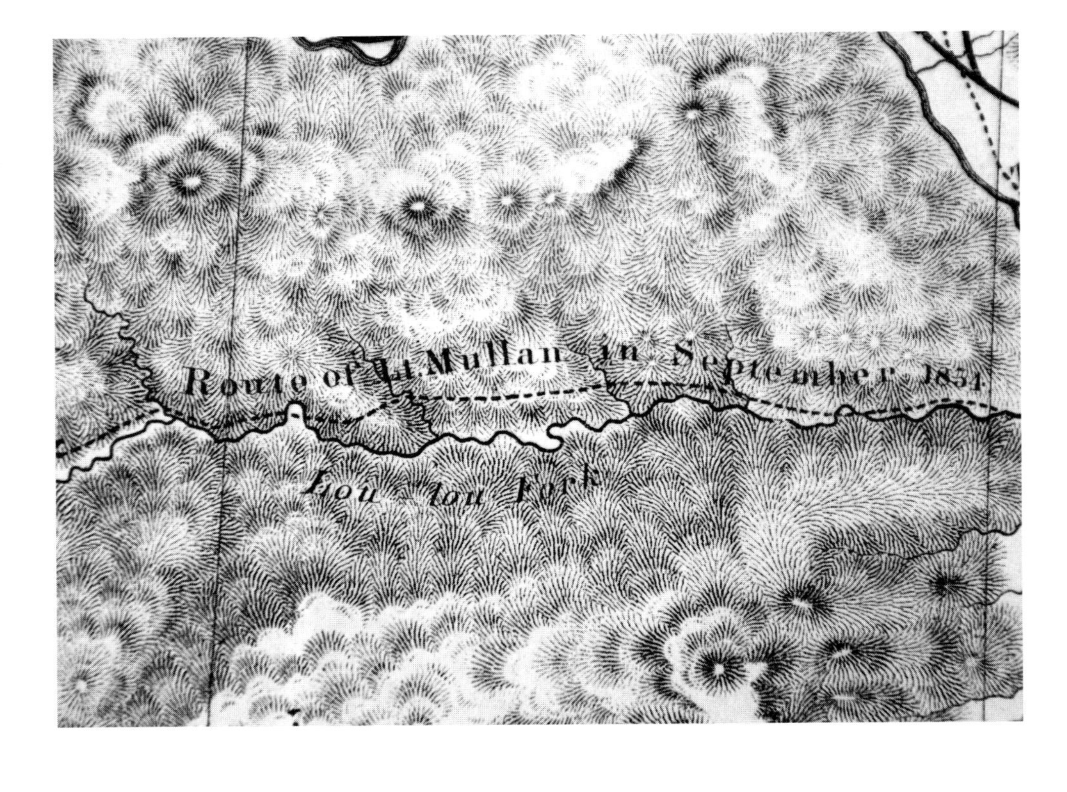
Route of Lt. Mullan in September 1854
Lou Lou Fork

PART III | imperial cartographies

17 Milk River to the Crossing of the Columbia River, a detail from Isaac Stevens's *Map of Railroad Survey of 47–49 Parallel* (Washington, DC: War Department. 1853–55).

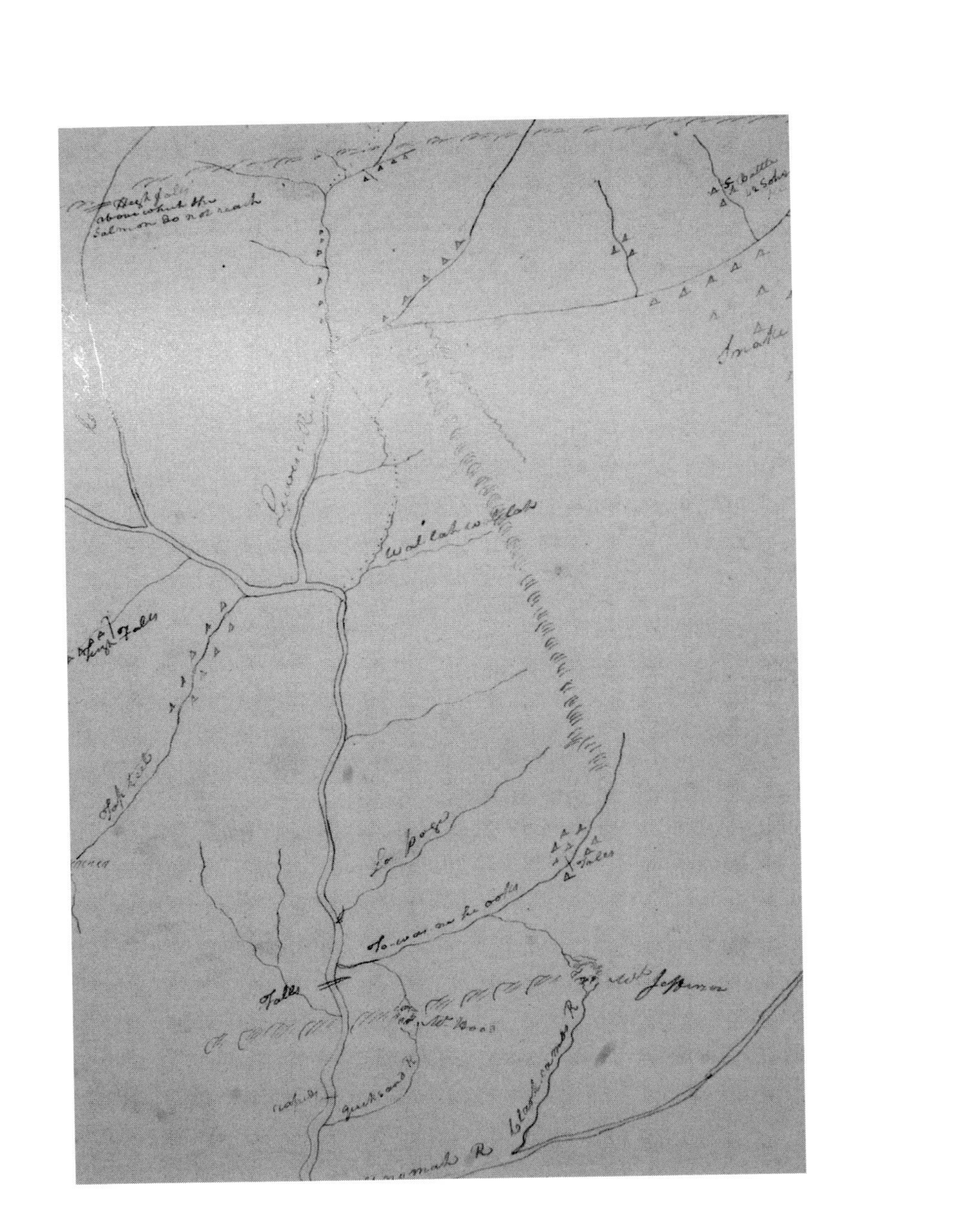
High falls above which the Salmon do not reach
Wal lah wollah
High Falls
La page
Falls
To-war-ne-he ooks
Falls
Mt Hood
Mt Jefferson
Quicksand R
Multnomah R

5 | Surveying the West: Lewis and Clark . . . and Others

The Lewis and Clark expedition is a hinged event that swings from the colonial to the imperial cartographies of the western United States. The expedition began when the United States was still only a fledgling nation restricted to the eastern seaboard. The expedition opened up the West to subsequent exploration that allowed the full emergence of a continental United States. Lewis and Clark inaugurated an exploratory practice with imperial reach for the new nation, as the United States staked a claim across the great continent and set about peopling the territory and commodifying the land. In the wake of Lewis and Clark, the Republic extended beyond the original thirteen colonies, west across the Mississippi to the Pacific Ocean, south to Mexico and north to Canada. This territorial annexation also involved the land's incorporation into national discourses of cartographic knowledge, scientific understanding, political control and economic calculation. The land required survey in all the subtle meanings of that term, including knowing, surveillance and control.

Before Lewis and Clark

On 20 June 1803 Thomas Jefferson sent a letter to Captain Meriwether Lewis outlining the goals of an expedition into the interior of the country. 'The object of your mission', it began, 'is to explore the Missouri

18 Cut Nose's sketch map, 1806.

river, & such principal stream of it, as, by it's course and communication with the waters of the Pacific ocean.'

Much has been written about this expedition that travelled from St Louis through the Rocky Mountains to the Pacific and back; it is now represented as a foundational expedition in the history of the United States. There is a vast body of documentation; here I will limit my remarks to noting the Amerindian cartographic contribution, which was significant. At least 30 Native American maps have been identified in the records of Lewis and Clark. And even before they left Washington, Lewis and Clark drew upon a large measure of Native American geographic knowledge encoded into the various maps that informed and guided their journey.[1]

James Mackay, a Scottish fur trader, and his Welsh assistant, John Thomas Evans, had explored the Missouri River in the 1780s and 1790s; they were sponsored by the Spanish government. They drew extensively on local informants. Planning to travel to the Pacific, Evans asked the Mandans to draw a map of the land to the west. The 1797 Mackay-Evans map entitled 'Sketch map of the Missouri river west of the Mandan villages, derived from Indian sources' shows the complex topography of the Rocky Mountains. The Mackay-Evans manuscript maps of the Missouri river provided very detailed information, and Lewis and Clark, while at Camp Dubois in Illinois during the winter of 1803–4, spoke with Mackay and had access to his maps.

Lewis and Clark

Lewis and Clark also took with them a manuscript map of the western United States by Nicholas King. Now in the Library of Congress, it drew upon a number of previous maps, including Peter Pond's and Aaron Arrowsmith's 1802 map of North America, Spanish military survey maps and maps by David Thompson. The Nicholas King 1803 map is the result of a complex borrowing from sources informed by Native American informants. The Ac ko mok ki map discussed in chapter Two directly influenced the Aaron Arrowsmith maps that King incorporated. The Lewis and Clark expedition used maps that drew upon myriad and complex cartographic encounters with Native Americans.

During the expedition Native American informants were exploited regularly. There is the well-documented role of the Shoshone woman Sacagawea who guided them at number of critical junctures, but Lewis and Clark relied upon the geographical knowledge of Native Americans throughout their journey. The continual reliance on informants is a constant thread in their narrative:

> The Indians inform that the yellowstone river is navigable for perogues and canoes nearly to it's source in the Rocky Mountains.[2]
> *Lewis, 26 April 1805*

> We were anxious now to meet with the Sosones or snake Indians as soon as possible in order to obtain information relative to the geography of the country.[3]
> *Lewis, 18 July 1805*

> The indians informed us that there was a good road which passed through the Columbia opposite to this village . . . they also informed us, that there were plenty of deer and antelopes on the road, with good water and grass. We knew that road in that direction if the country would permit it would shorten our rout at least 80 miles. The indians also informed us that the country was level and the road good.[4]
> *Lewis, April 1806*

> Indians informed us that we could not cross the mountains until the full of the next moon; or about the 1st of July. if we attempted it sooner our horses would be three days without eating.[5] *Clark, 8 June 1806*

In their journals Lewis and Clark also often noted their reliance on the mapmaking of the Native Americans: 'This War Chiefe gave us a Chart in his Way of the Missourie' (*Clark, 16 January*).[6] 'In his Way' probably meant drawn on the ground. Here is an example:

> I now prevailed upon the Chief to instruct me with rispect to the geography of the country. This he undertook very cheerfully, by delineating the river on the ground . . . he drew

> the river on which we now are (Lemhi River) to which he placed two branches just above us . . . he next made it discharge itself into a large river which flowed from the S.W. abut ten miles below us (Salmon River), then continued this joint stream in the same direction of this valley . . . for one days march and then encircled it to the West for 2 more days march. Here he placed a number of heaps of sand on each side which he informed me represented the vast mountains of rock . . .[7]
> *Lewis, 20 August 1805*

In this encounter the chief went on to note the qualities of the river, the nature of the riverbed and the speed of stream flow. It was an extraordinarily complex itinerary map, as well as a sophisticated topographic map.

Maps were also drawn on animal skin: 'I got the Twisted hare to draw the river from his Camp down which he did with great Cherfullness on a white Elk skin' (*Clark, 22 Sept, 1805*).[8] They were also transcribed onto paper, especially by William Clark. We have a rich trove of such Native American maps, including Shehek-Shote's 1805 map and maps done in 1806 (illus. 18) as well as maps by members of what Clark noted as the Chopunnish (Nez Perce), Clatsop and Skaddot (Klickitat) tribes.[9] Illustration 18, for example, is a map drawn by Clark based on information supplied by a member of the Nez Perce tribe, Cut Nose. The map depicts the Columbia River up to the Snake River. It is a sophisticated depiction. Distance is rendered in space–time terms; easier sections to traverse are more compressed on the map compared to more difficult sections. Illustration 19 is a detail of the map drawn by Clark based on information from a number of indigenous informants from the Klickitat, Nez Perce and Clackamas tribes. The river system is presented as a geometric system indicating a complex understanding of spatial representation. Only recently a Native American map was found in the Missouri Historical Society: it was made by Nez Perce chiefs and given to William Clark; it showed a short and safe route back through the Rocky Mountains.[10]

The success of the Lewis and Clark expedition, so often viewed as a triumph of white explorers, with a more recent nod toward the heroic role of Sacagawea, is due in no small part to a complex

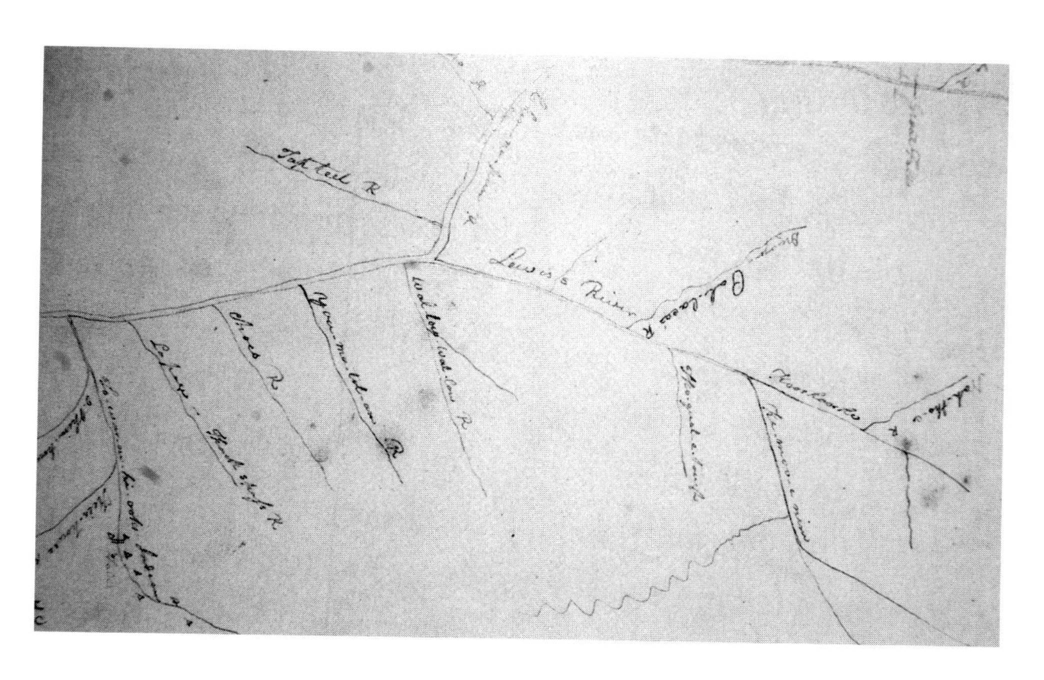

19 Skaddot's map drawn by Clark, 1806.

cartographic encounter in which the expedition set out with maps shaped directly and indirectly by Native Americans. An argument could be made that Sacagawea was so vital to the expedition because the encounter was so complex. She was a vital element in the subtle and myriad transmissions of geographical information and cartographic representation. The successful course of the expedition was dependent on the testimonies of Native American informants, guides and mapmakers.

On his return east Clark compiled a manuscript map entitled *A Map of Part of the Continent of North America* which was a new depiction of the West that relied heavily on cartographic encounters. The map was forwarded to Samuel Lewis in Philadelphia, who printed it in 1814 (illus. 20). *The Lewis and Clark Track Map*, as it is often referred to, depicts the complex topography of the Rocky Mountains and the sinuous course of many of its rivers: it was a landmark in western cartography that influenced the perception of the region for generations. And here we come to the destructive twist at the heart of such cartographic encounters: the 1814 map provided the geographical knowledge to promote and aid further western expansions. Widely

distributed, it presented a more accurate rendering of land than was previously known or available. Connected rivers and wide-open spaces presented a land waiting for exploitation. Very dependent on Native American sources, the Track Map became *the* map that led to the unravelling of the Native American hold on territory. And, in another inevitable irony, a map so reliant on Native American sources largely ignored Native Americans. Tribal names were cited on the map (see illus. 21), but no mention was made of Native American trade routes, sacred places and settlements. This map erased major elements of a long-standing human presence. Clark, who had relied so heavily on so many Native American mapmakers, informants and

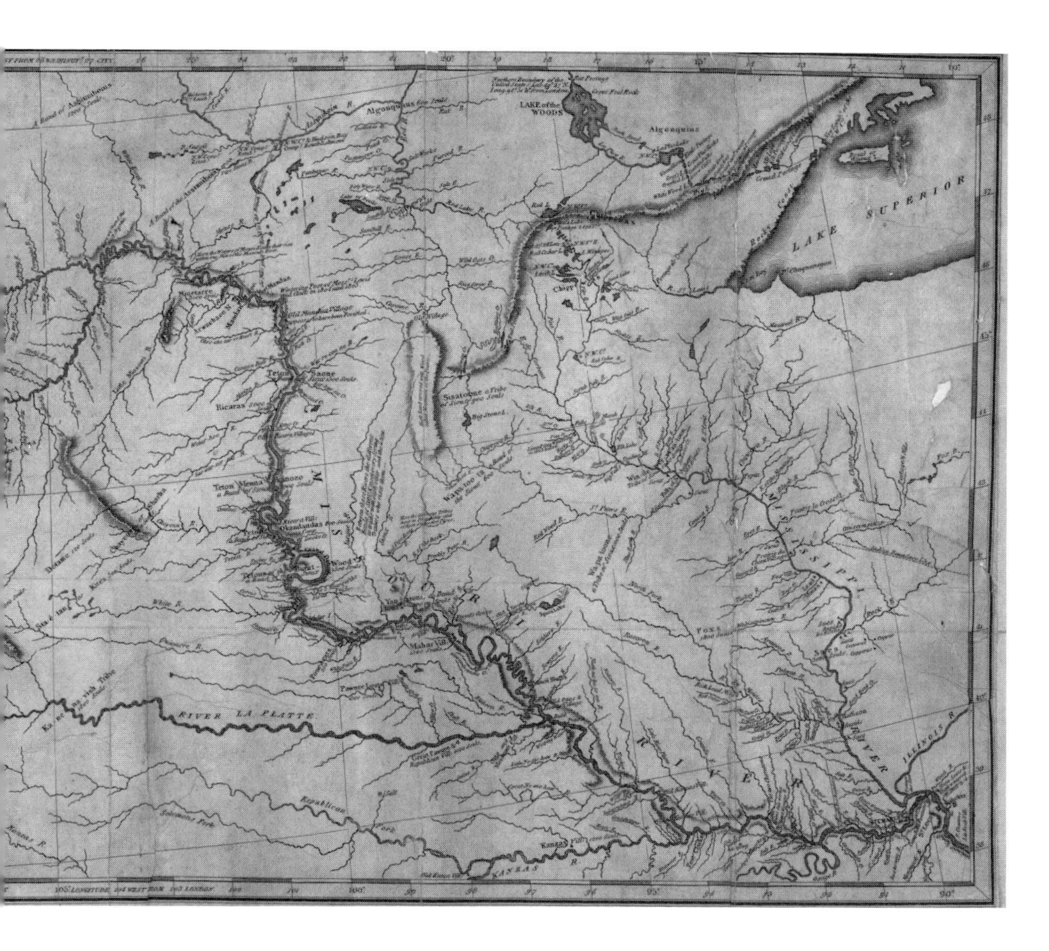

20 The *Map of Lewis and Clark's Track Across the Western Portion of North America* (Philadelphia, 1814).

guides, co-authored a map that not only denied much of their presence initially but ultimately led to their annihilation. Just as Clark did in the map so he did in his post-expedition career. As a federal official in his later life he oversaw the removal of 100,000 Indians. From 1808 to 1838 during Clark's tenure as the most important federal official in the West, and under his direct influence, the Cherokee, Chickasaw, Choctaw, Delaware, Kickapoo, Miami, Osage, Ottowa, Potrowatomi, Quapaw, Seneca, Shawnee and Wynadot were all forcibly removed from their lands.[11]

A quotation from Carl Becker inscribed on the US Capitol building reads 'To venture into the wilderness one must see it, not as it is,

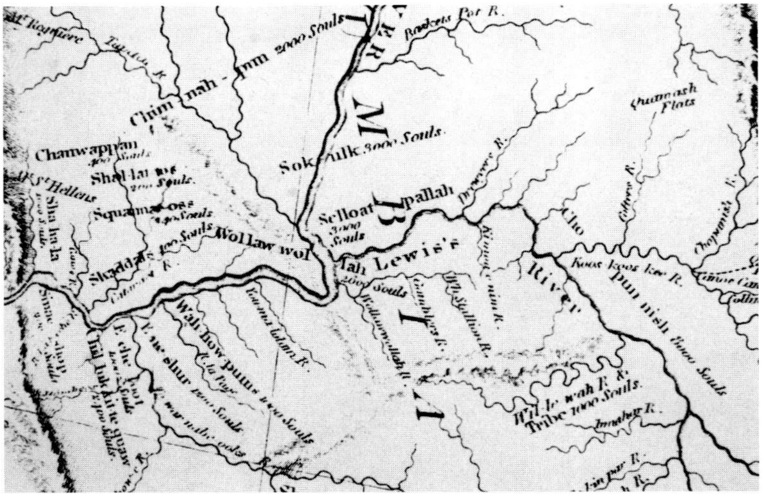

21 Detail from *The Lewis and Clark Track Map*.

but as it will be.' Clark ventured into the wilderness and recorded it through an extensive and intensive reliance on Native American sources. And through this cartographic encounter the reality of an already peopled land was transformed into an imperial United States virtually emptied of its indigenous population.

6 | Expedition into the 'Desert'

There were many more explorations and surveys, frequently published, after Lewis and Clark. Between 1800 and 1861 there were approximately 80 military-scientific explorations, ranging in time from Sibley's 1803 *Exploration of the Red River* to Raynold's 1859 *Exploration of the Yellowstone*. These enterprises bounded the expanding nation from the *United States and Mexican Boundary Survey* of 1848 to the 1857 *Northwest Boundary Survey*. Not restricted to the US, they spread around the world, from the Rodgers and Ringold *North Pacific Exploring Expedition* of 1853 to the 1851 Herndon and Gibbon *Explorations of the Valley of the Amazon*. Some were mostly civilian and largely private, such as Nicollet's 1836 *Exploration of The Upper Mississippi*, others predominantly military and federally backed, such as the *US Exploring Expedition of 1838–42*. Some were general in scope, such as the 1859 Macomb *Expedition in New Mexico, Utah and Colorado*, some very focused, including Schoolcraft and Allen's *1832 Expeditions to the Source of The Mississippi*, Owen's 1847 *Survey of the Chippewa Land District of Wisconsin* and the 1849 *US Survey of the Creek Boundary Line*. Some became famous and entered national legend, such as Fremont's expeditions of 1842, 1843 and 1845, others still remain relatively well known, such as the *Pacific Railroad Surveys* of 1853, and many now languish in obscurity. Who but specialized historians know anything of Featherstonehaugh's 1836 *Report of a Geological Reconnaissance made in 1835 . . . to the Coteau des Prairie*?

Despite the many differences between the various expeditions they share three important characteristics. First, they generated knowledge of territory. This took the form of written reports, maps, specimens and samples. To collect and represent this information, the larger expeditions were filled with scientists and artists who measured and collated, drew and painted. The results were recorded in reports and represented in maps and lithographs that soon entered the public domain. The words and images quickly percolated back east to amplify understanding of the expanding United States. The maps, drawings, collections and specimens formed the basis for the national archives of accumulated scientific knowledge. The survey reports and images shaped the scientific understanding and public perception of the history and geography of the nation.

Second, the various accounts participated in an intellectual framework shaped by the emergence of an international scientific community closely allied to imperial claims. This scientific–imperial complex developed in the eighteenth century and flourished in the nineteenth century. The iconic heroes were James Cook and Alexander von Humboldt. The journeys and expeditions of Captain James Cook (1728–1779) were well known. He led three expeditions to the Pacific Ocean, Antarctica and the Arctic, recording new geographies and mapping new realms. Cook was the most visible representative of a new scientific discourse that stressed fieldwork, careful measurement and the mapping of new worlds. He charted Hawaii as well as the coastlines of New Zealand and Australia. His journals were turned into official reports and published to wide acclaim in 1774. Other members of his expeditions, such as the socially connected Sir Joseph Banks, also produced popular narratives. The first biography of Cook appeared in 1788 and he became a revered figure for enlightenment thinkers in colonial America and the early Republic. The American navy afforded him safe passage during the revolutionary war with the British. As other nations sought to emulate the British nexus of science and imperialism, so other explorers modelled themselves on Cook. Jean-François de Galoup, Comte de La Pérouse (1741–1788) was a French explorer and naval officer who much admired Cook. In 1785 the king of France made La Pérouse the leader of a two-ship expedition to the Pacific Ocean, which included scientists, naturalists, a physicist, three draughtsmen and a mathematician. By 1786 the ships had reached Hawaii and went on to map the North American west coast from Alaska to California before sailing to Australia. Although all members of the La Pérouse expedition perished during a storm near the Solomon Islands, an atlas of the expedition's world travels was soon published. An English edition appeared in the 1790s.

Alexander von Humboldt (1769–1859) was an influential figure in the development of US scientific fieldwork. From 1782 to 1792 he studied at universities in Frankfurt, Gottingen, Hamburg and Freiberg. Widely read, he had interests in economics, geology, botany and mining; independently wealthy, he was able to travel. With encouragement and permission from King Carlos IV Humboldt set out for Spanish America. He arrived in Venezuela in 1799 and by 1800 had travelled down the Orinoco. For the next four years he and his companion Aimé Bonpland travelled around South America climbing mountains, making maps, documenting the physical world. In 1804 they sailed to the United States where the nascent scientific community warmly received them. Humboldt stayed with Jefferson at Monticello for four months.

Humboldt made meticulous observations, but he was also looking for general trends. From his observations of mountain vegetation, for example, he generated a model of vegetation type varying with altitude just as it did with latitude. He believed in careful measurement but also searching for the interrelations between diverse phenomena. A broadly trained physical scientist, he was also aware of the connections between human and physical systems. Humboldt was one of the first modern geographers, and his form of scientific expedition, analysis and writing became a model for US western exploration.[1]

Third, the various accounts, charts and maps emerging from expeditions relied in varying degrees on cartographic encounters with indigenous peoples. While the first two characteristics have been well documented in the literature on scientific expeditions, this last point is weakly developed. It is as if the rational emphasis on science and nationalism, measurement and appropriation ignores, downplays or miscasts the role played by the indigenous peoples. In this and the next two chapters I will look again at these western expeditions for further evidence of cartographic encounters. In this chapter I will focus on the Long Expedition.

Taking its name from its leader Major Stephen Long this expedition began in Pittsburgh in early May 1819 as part of a larger enterprise to establish a fort, control the Indians and drive out British fur traders. Under General Henry Rice Atkinson over a thousand men were sent in five steamboats up the Missouri. The Long Expedition members were in the sixth steamboat. Long was given permission to explore the Great Plains between the Missouri River and the Rocky Mountains. His charge from the Secretary of War, and one that could have been written for all subsequent American explorations, was 'to acquire as thorough an accurate knowledge as may be practicable of portion of the country, which is daily becoming more interesting, but which is as yet imperfectly understood . . . you will permit nothing worthy of notice to escape your attention'. The Secretary also suggested that Jefferson's order for the Lewis and Clark expedition provided a good guide.[2] The term 'accuracy' in the charge was not accidental. Emphasis was placed on measurement. The calibration of altitude, temperature, latitude and longitude were primary objectives, necessary in order to plot the land.

In order to let 'nothing worthy to escape your attention' a number of scientists and artists were sent on the Long expedition. Although led by an army officer and ostensibly a military expedition, it also comprised civilian men of science and art. Long asked the American Philosophical Society in Philadelphia to suggest likely people for the trip. The expedition initially included a zoologist (Thomas Say), a physician and botanist (Dr William Baldwin), a geologist (Edward Augustus Jessup), an official artist (Samuel Seymour) and an assistant naturalist (Titian Ramsay Peale). Two of the three army officers were topographical engineers while Captain Bell was assigned as a journalist. Dr Edwin James replaced Baldwin and Jessup in 1820. The final party also included guides and interpreters.

The Long party travelled along the North, and then the South, Platte River, climbed Pike's Peak and then continued south. One group, led by Bell, went east along the Arkansas River while another group, led by Long, went down the Canadian River. The two groups met up at Fort Smith in 1820.

The expedition suffered from the usual problems of frustration and error. An early exploration, it did not have the proper scientific instruments to measure the height of Pike's Peak. When almost within reach of Fort Smith, three soldiers in Bell's party deserted, taking saddlebags containing many of the scientific notebooks. Long mistakenly thought that the Canadian River was the Red River until it was too late to retrace his steps. All in all, a standard exploration.

'You will enter in your journal every thing interesting in relation to soil, face of the country, water courses . . . ' The nineteenth-century explorations share a common theme of journal production. The standard format of the journal is a beginning that states the official orders and lists the members of the party with their responsibilities. It then takes a diary form with entries written up each day. This gives the journals an immediacy and intimacy as if the reader was accompanying the expeditions through their daily endeavours. In reality the journal entries started as rough notes, written up often days and even weeks after the actual events and then given a polished finish sometimes months, and in some cases years, later. The journals seem, at first blush, a simple recording of the day's events, but they are a complex form of representation. Two main and competing themes can be identified: the description of an occupied space peopled by indigenous peoples and the depiction of an empty space that awaits the full unfolding of the Republic. The journals represent the paradoxical geography of a land at once occupied yet empty.

The Long Expedition, as many of the other expeditions, produced a number of texts. A narrative report, *An Account of An Expedition from Pittsburgh to the Rocky Mountains*, was first published in 1823. It was written by Edwin James and drew upon, as it noted in the full title, 'the Notes of Major Long, Mr. T. Say, and other Gentlemen of the Exploring Party'. The Long Expedition report is unusual in that it admits its complex production and it reads more like a composite report than a personal diary. We also have a more backstage view of the report and the expedition. The principal compiler, Edwin James, also wrote letters to his brother.[3] In December 1819 it is clear from these letters that James was at a loose end, frustrated, indebted and disappointed with his lack of progress in the world. In early February 1820 he reports joyfully that his application for an appointment in the 'western expedition' has been successful. His prospects suddenly improve. James's letters to his brother from the expedition provide geological observations (his brother was studying geology) and comments on the Indians as well as personal remarks. On October 26 he reports 'I have had an attack of fever and ague . . . I am not likely to recover.' He went on to complain about the lack of time, 'I am full of complainings and bitterness against major Long. I have been allowed neither time to examine and collect or means to transport plants and minerals. We have been hurried through the country.' James's letters provide a more intimate portrait of an expedition, full of the illness, fatigue and personal animosities that surely encumber all expeditions.

The report was compiled and written by James on his return, when, as he stated in a letter to his brother, he was 'comfortably planted in Philadelphia'. It is an educated travelogue that contains extensive information on the native peoples. Large chunks of the text are taken up with detailed descriptions of the different tribes, including the Konzas, Otoes, Missouries, Ioways and Pawnees: it is a description of an occupied space, a settled land, full of complex societies. Native Americans figure hugely in the report. Their manners, customs, dress and economy are given detailed descriptions. The Long Expedition is, despite later depictions of an exploration into the wilderness, a journey though a populated landscape. On 4 October 1820,

> At ten o'clock, the hour appointed for the council, the Indians headed by their chiefs arrives; and after shaking us all by the hand took their seats. There were about one hundred Ottoes, seventy Missouries and fifty or sixty Ioways. They arranged themselves agreeably to their tribes, on puncheon benches which had been prepared for them, and which described a semicircle.[4]

We also have a visual record of this encounter. The expedition's official artist Samuel Seymour depicted the event in one of his paintings that was reproduced as an engraving in the *Account* (illus. 22). And just as we have James's letters to provide a background perspective, so we have Peale's drawings. Although not the official artist, Titian Ramsay Peale, an accomplished painter from a family of artists, also made artistic renditions of the journey. He drew 122 sketches, some of which provided the basis for watercolours that were subsequently displayed at the Philadelphia Museum. I examined five of his sketchbooks that are held at the Yale University Art Gallery. The first four are dated 1820 and show Native Americans, rivers, birds and

22 Samuel Seymour, *Major Long Holding a Council with the Oto (Missouri) Indians*, 1822-23, watercolour.

deer. They are delicate portraits, more intimate than grand, that suggest an English pastoralism rather than the romanticism that so dominated later depictions of the West. The fifth book is dated 1823 and was produced in Philadelphia. The drawings, an eclectic mix of topography as well as botanical and zoological specimens, stand out as bright flashbacks, miniatures of a remembered West.[5]

Maps were among the important texts produced by the Expedition. Long produced a large pen-and-ink manuscript map, now in the US National Archives, in 1820: it shows a large area from Washington, DC to the Rockies with the rivers Platte and Canadian shown in great detail. This manuscript map was the basis for one published in 1822 in *Maps and Plates*, published to accompany the *Account*. The published map is in two sections, Eastern and Western, and was soon reproduced in commercial publications. The same year the map was published in a Carey and Lea Atlas as the *Geographical, Statistical and Historical Map of Arkansas Territory*. The next year it appeared in Tanner's *New American Atlas*. Very quickly Long's map became part of the accepted view of the West. In his manuscript map Long had placed the words *Great Desert* to mark the southwestern plains. This designation also appeared in the Carey and Lea *Atlas*. In the official *Maps and Plates* of 1822 the term appears as *Great American Desert*. Although there is a great deal of detail on this map (as shown in illus. 23), it shows rivers, routes,

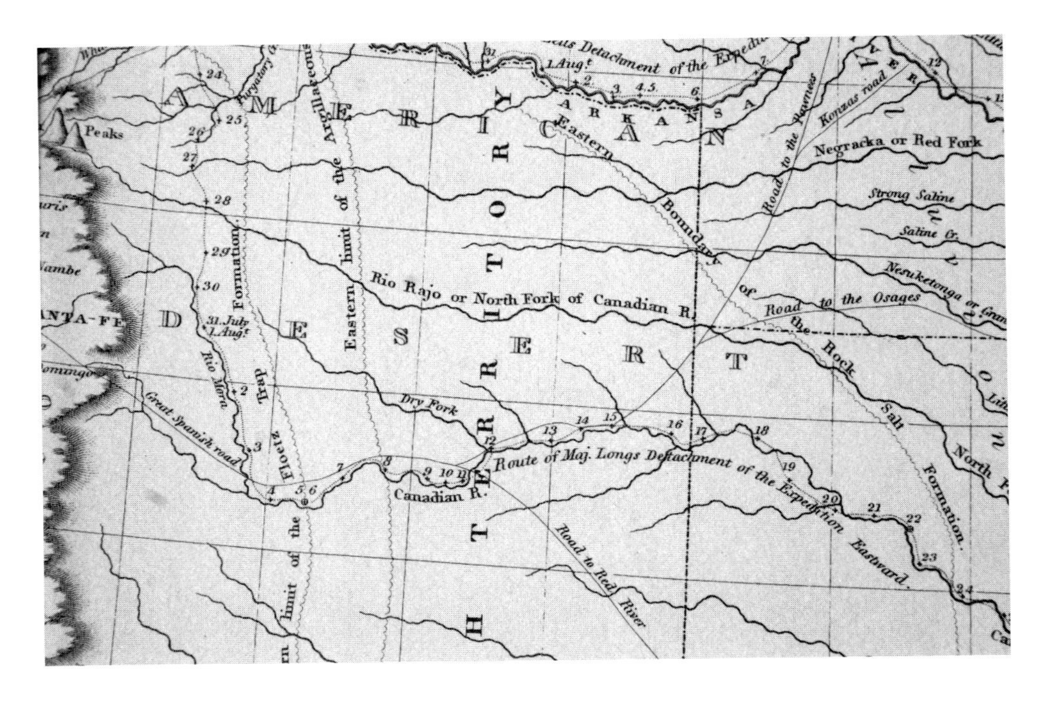

23 Detail from Stephen H. Long's *Geographical, Statistical and Historical Map of Arkansas Territory* (1822).

mountains and the path followed by the Long expedition in often confusing profusion, and the word 'Desert', from the full listing 'Great American Desert', can be seen. Meant in part to imply a buffer zone between competing empires, the term 'Desert' inhibited white emigration to the area for decades. The designation of desert is interesting and also puzzling. While many definitions of the term were in common usage at the time, it was associated with being uninhabited and desolate. And yet, as the *Narrative* richly details, it was a landscape populated by a variety of different tribes. Long's designation on the map and the *Narrative* highlight the paradoxical geography of an empty and occupied space in the same territory.

This paradox is embodied in some of the other Seymour illustrations that accompanied the *Account*. Illustration 24, for example, depicts the sweeping grandeur of the western landscape across the vast animal grazing plains to the Rocky Mountains. But note that there is a Native American presence in the left-hand corner of the map. And as befitting an expedition through an occupied space, the local inhabitants are also considered worthy of depiction.

Long's map is regarded as a landmark cartographic achievement. Here is a recent and standard description: 'Long's expedition was the first scientific survey of the southern plain, and, with the aid of a few basic instruments, his map added much to the geographical knowledge of the area west of the Mississippi River.'[6] Nowhere is any credit given to the Native American contribution, but if we read the *Account* with sensitivity, we can easily find it. It is full of cartographic encounters. During the first winter at Engineer Cantonment the *Account* identifies the cartographic syntax of sign language used between the indigenous people and members of the expedition. A river, for example, was a hand making a scoop to the mouth, then the drawing of a horizontal line. Upstanding fingers of the left hand identified mountains with the right-hand index finger forming serpentine flows. The report documents frequent encounters. On 9 August 1820,

> To our inquiries concerning the river, they answered without hesitation, that it was the Red River; that at the distance of ten days traveling, in the manner of Indians with lodges (about one hundred miles) we should meet with permanent village of the Pawnee Piques; that a large band of Camancias were hunting on the river below; whom we should fall in with two or three days. Having described to them the route we had pursued, and the great and frequented road on which we had travelled, they said that when we were at the point where that road first crosses the river, we were three days ride from Santa Fe, which was situated behind a low and distant range of hills.[7]

The Expedition made their way through the territory along Native American trails. Here is a selective reading of the *Account* for 19 August to 31 August 1820. 'At a short distance from our place of encampment we passed an Indian camp . . . Well worn traces or paths lead in various directions from this spot.'[8] They take a route along the river but heavy forest makes it difficult to follow the river. The next day they get caught in a heavy storm: 'The ravines were muddy, and their banks slippery in consequence of the rain; we had, however, the good fortune to fall upon an Indian trace, which complied with our proper direction, and which indicated the best points at which these

24 Samuel Seymour, 'Distant View of the Rocky Mountains', frontispiece to Ewin James's *Account of an Expedition from Pittsburgh to the Rocky Mountains . . . Under the Command of M. Stephen H. Long* (Philadelphia, 1823).

gullies might be passed.'[9] The next day, 'The Indian trace was again discovered, and pursued about nine miles to the dining place at noon'.[10]

They made their way through the country by following Indian paths and camping at Indian camping sites. On 28 August they report,

> In the afternoon, having descended to the river, we again laboured through the difficulties of dense underwood which such productive soils usually present, until toward evening, when we had the happiness to see a well worn Indian path, which had been interrupted by the river, and now took a direction towards our left. Wishing to pursue this route as well for the facility of travelling, as with the hope of soon arriving at some Indian town . . . with little hesitation, therefore we struck into the path.[11]

And the next day, 'We pursued the Indian path a considerable distance.'[12] And the day after that,

> Another Indian path was now discovered, which by its direction seemed to comply with our proper course. It led us to recross the ravine with its most luxuriant growth of trees, bushes and weeds. On emerging from this intricate maze . . . the appearance of well-worn pathways, inspire us with a renewed expectation of soon meeting with human beings, and of arriving at some permanent Indian village.[13]

Although the official map traces their movement through the 'Great American Desert' (illus. 23), for hundreds of miles the Long Expedition followed Indian trails that allowed them to cross rain-washed gullies, pass through dense underwood and emerge safely from an 'intricate maze' of trees, bushes and weeds. Long may have designated the region a 'desert', but he travelled through it along well-worn pathways, traces and trails. Long and his group were less explorers, more travellers following well-trodden routes.

7 | Fremont and Tah-Kai-Buhl

John Charles Fremont is a well-known figure in the history of the American West (illus. 25). He gained fame for leading five expeditions to explore the West between 1842 and 1854. With his wife's help and powerful family connections he parlayed his expeditions' good publicity into a career as a US senator and an 1856 Republican Party presidential candidacy. Throughout the campaign the emblematic image of 'Fremont the Pathfinder' was invoked with frequency.

But what of the other name in the chapter title, Tah-Kai-Buhl (illus. 26)? There seems little connection between them; a reading of two recent Fremont biographies and the Fremont entry in the authoritative *Encyclopedia of the American West* makes no mention of this name.[1] Fremont the Pathfinder and who? Yet, as I will show, it is more accurate to describe Tah-Kai-Buhl as the Pathfinder. And this particular encounter stands for many more Fremonts and Tah-Kai-Buhls.

To tell the story of this embodied cartographic encounter, we must note the intersection of the two characters both in some detail and also from a wide angle. While their individual stories are interesting inherently, they also represent much broader cases of mapping collaborations in the military exploration of the West. Let us begin with Fremont. He was born in 1813 in Savannah, Georgia. His mother ran away from her first husband to be with Fremont's father, a French émigré. From the age of five he was raised in Charleston. Fremont

25 'Major-General John C. Fremont', undated lithograph.

26 James W. Abert, *Tah-kai-buhl. A Kioway*, 1846, watercolour tipped into his own copy of *Message from the President . . . Communicating a Report of an Expedition Led by Lieutenant Abert, on the Upper Arkansas and through the Country of the Camanche Indians, in the Fall of the Year 1845* (Washington, DC, 1846).

certainly had a knack for making important connections that advanced his career. The first was Charleston-born Joel Poinsett, a board member of the college that Fremont attended and member of the same local church. Poinsett was an important politician who at various times served in the US House of Representatives, and was Secretary of War as well as the first US Minister to Mexico. His name lives on in the name of the popular Christmas flower that he sent home after a visit to southern Mexico. With Poinsett's influence, Fremont was appointed to a position aboard the USS *Natchez* teaching mathematics to midshipmen. On his return he worked – again through the influence of his mentor – on surveys of projected railways through Indian country. When Poinsett became Secretary of State for War in the Van Buren administration, he arranged for Fremont to be commissioned as Second Lieutenant into the US Corps of Topographical Engineers in 1837.

The Corps was the latest military incarnation of surveying national land. The United States obtained a huge area west of the Mississippi that doubled the size of its territory with the Louisiana Purchase in 1803. A magnificent addition comprising vast plains filled with seemingly endless herds of buffaloes, high snow-capped mountains, huge inland basins, flowing rivers and deep lakes, the Louisiana Purchase presented the United States with a generous helping of nature's bounty. Yet such richness and wonder could only be imagined because it lay beyond the knowing of the Republic; it was, as the early and even some later maps of the territory would note, 'unexplored country' (illus. 27). The transformation from unexplored to explored involved a large number of expeditions that found, surveyed and mapped.

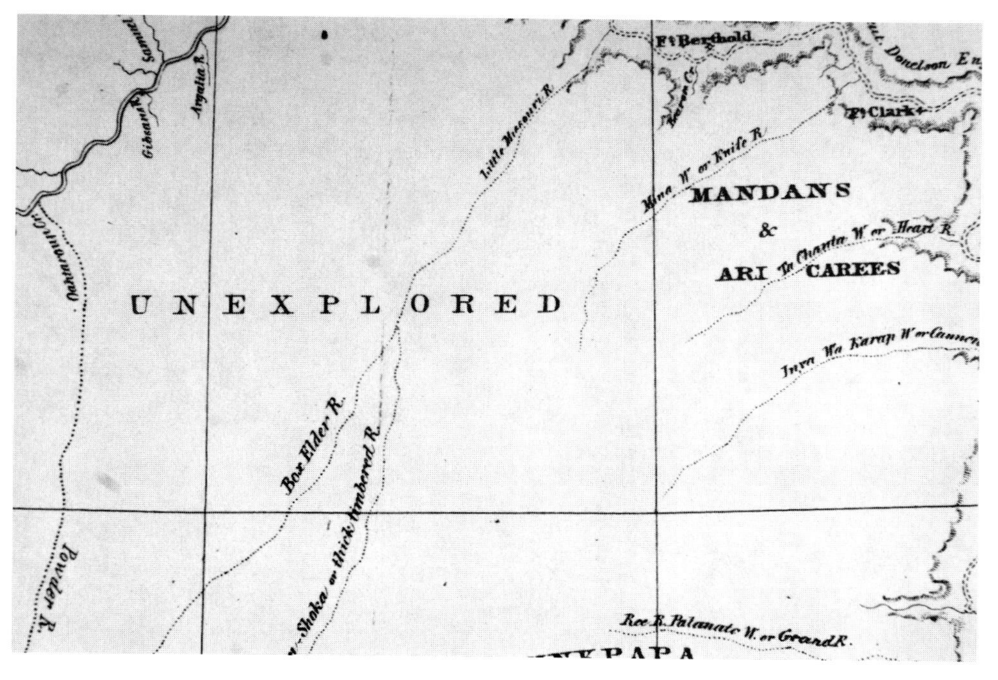

27 Detail from *Map of Lt. Warren's Report of Military Reconnaissances in the Dacota Country*, 1855.

Much has been written of the military exploration of the West.[2] A number of assumptions underlie the popular conception that the West was a near empty wilderness discovered by white explorers. The overriding thrust of the narrative is of the white exploration of a wilderness, albeit nuanced with an awareness that Native Americans were involved. A more accurate narrative would tell of a cartographic collaboration with Native Americans sometimes willingly, often unwillingly, using their skills, knowledge and resources to show the way. The so-called explorers were not so much discoverers as partners in collaboration. And there was no true wilderness, despite the terminology of some.[3] The West was inhabited, cared for, cherished, a source of sustenance as well as meaning for a myriad of peoples, which collectively we now call Native American. These peoples varied in speech, allegiance, beliefs and material culture; theirs were not static societies as they had been subject to many changes, including the arrival of the horse as well as the ongoing dynamics of intricate trading arrangements and delicate treaties and pacts. The Native Americans had been working this land through agricultural practices, hunting and settlement for over 15,000 years. The United States obtained a vast land full of physical wonder and complex societies, more garden than wilderness.

Various expeditions were sent out to map this new land in the course of territorial appropriation that brought the land and its peoples under political control, scientific understanding and national recognition. As an enterprise of the Corps of Discovery Lewis and Clark were dispatched by Jefferson. Later, the Topographical Bureau was established in the Army's Corps of Engineers in 1813, a small unit rarely with more than 10 members. In 1829 John James Abert was made its head. A graduate of West Point, Abert had served with the various army topographical units since 1814. He worked on the geodetic surveys of the Atlantic coast and the survey of the Chesapeake and Ohio Canal. Under Abert the Bureau expanded, and in 1838 it was re-established as the US Corps of Topographical Engineers, on equal footing with the Corps of Engineers. From 1838 until 1863 when it was folded into the Corps of Engineers, this elite group – under Abert's deft leadership for all but two years – played a major part in the mapping the West. Members included William Emory, Andrew Humphreys, Lorenzo Sitgreaves, Gouverneur Warren and Amiel Whipple as well as Fremont. Projects included the mapping of the Great Lakes and Mississippi River basin, boundary surveys in the Southwest and Northwest and the transcontinental railroad surveys. At its peak the Corps had 72 officers, most of them graduates of West Point. It was well funded and staffed by professionals with a strong meritocratic rather than preferment culture of promotion.

The Corps was actively involved in the development of science in the United States. Abert believed in surveying for military and commercial interests as well as for the collection of scientific data. In those times the distinction between the scientific and military communities was less apparent than today and army officers were often part of a broader scientific community. For example, in 1854, Captain William Hammond on the frontier at Fort Riley in Kansas Territory was elected into the Academy of Natural Sciences as a corresponding member. In his letters to civilian scientists in Philadelphia he writes of his scientific work with a microscope and of his collection of animal specimens. The correspondence highlights the ongoing exchange between scientists in the city and army officers on the frontier. The officers in the field could provide data and observations in return for equipment and news of the latest scientific developments.[4]

Abert was a member of various scientific societies including the Geographical Society of Paris, and was involved with many US scientists: Audubon named a squirrel after him. In 1832 he was elected to the American Philosophical Society, the foremost scientific society of its day in the United States. The nomination noted 'the published essays of Colonel Abert on subjects connected with the Natural Sciences, his high standing as an officer of the honorable Corps, in which he has so often distinguished himself, his acknowledged devotion to the cause of science'. The Corps included scientists and savants in its explorations, and the reports and specimens from many of the Corps' expeditions and explorations became part of the nation's scientific and anthropological archives.

The Nicollet survey

Fremont was commissioned into this elite scientific surveying at the beginning of the Corps' existence. Through the promptings of Poinsett, his first assignment was to accompany the expedition of the Frenchman Joseph Nicollet into the Upper Mississippi and Missouri rivers.

Joseph Nicolas Nicollet (1786–1843) was born in Savoy in eastern France. An accomplished violinist, he tutored young children before he went to Paris to train for a teaching career. He developed an interest in astronomy and mathematics, wrote scholarly papers and was noticed by Laplace who got him a job at the observatory of the Military School. When only 21 years of age he discovered a comet. After losing money in the Bourse in 1831 he emigrated to the US in 1832 for, he wrote, 'the purpose of making scientific tours'.[5] His goal was to map the Mississippi River Valley, much of its upper reaches then unknown to science. He befriended the powerful and rich Choteau trading family, which financed his expedition to look for the source of the Mississippi. He was also encouraged by the War Department which, eager for information, provided letters of protection and hospitality for army frontier posts.

28 Detail from J. N. Nicollet's *Map of the Hydrographical Basin of the Upper Mississippi River* (1843).

In 1837 Abert wrote to Nicollet requesting the geographical information that he had collected. Nicollet asked for $2,500 for a map of the region, with $1,000 as a down payment. Nicollet also persuaded Poinsett and Abert of the need for more information on the upper Mississippi region. With Poinsett's approval, Abert then hired Nicollet to lead two surveys. The first, in 1838, was from present-day Minneapolis west to eastern South Dakota. The second, much longer trip, the following year, left from St Louis, travelling through much of the Dakotas. Fremont was assigned to accompany Nicollet and spent two summers in the field with him.

Nicollet was the second major influence on Fremont. While Poinsett was a political figure, Nicollet was an important teacher. Fremont gained sophisticated scientific instruction in science and surveying. Nicollet prized careful measurement; he was one of the first to use a barometer to gain accurate altitude readings. The Nicollet explorations through the prairie territory resulted in two major government publications: *The Map of the Hydrographical Basin of the Upper Mississippi River* in 1843 (illus. 29) and an accompanying *Report* published two years later.[6]

The *Map* is widely regarded as one of the earliest accurate maps of the prairie country of the upper Mississippi: it shows altitude and the river's course for over one thousand miles. G. K. Warren – we will discuss him later – described it as one of the great contributions to American geography. The *Report* is a relatively small book with no illustrations apart from a smaller scale version of the *Map*. It reads

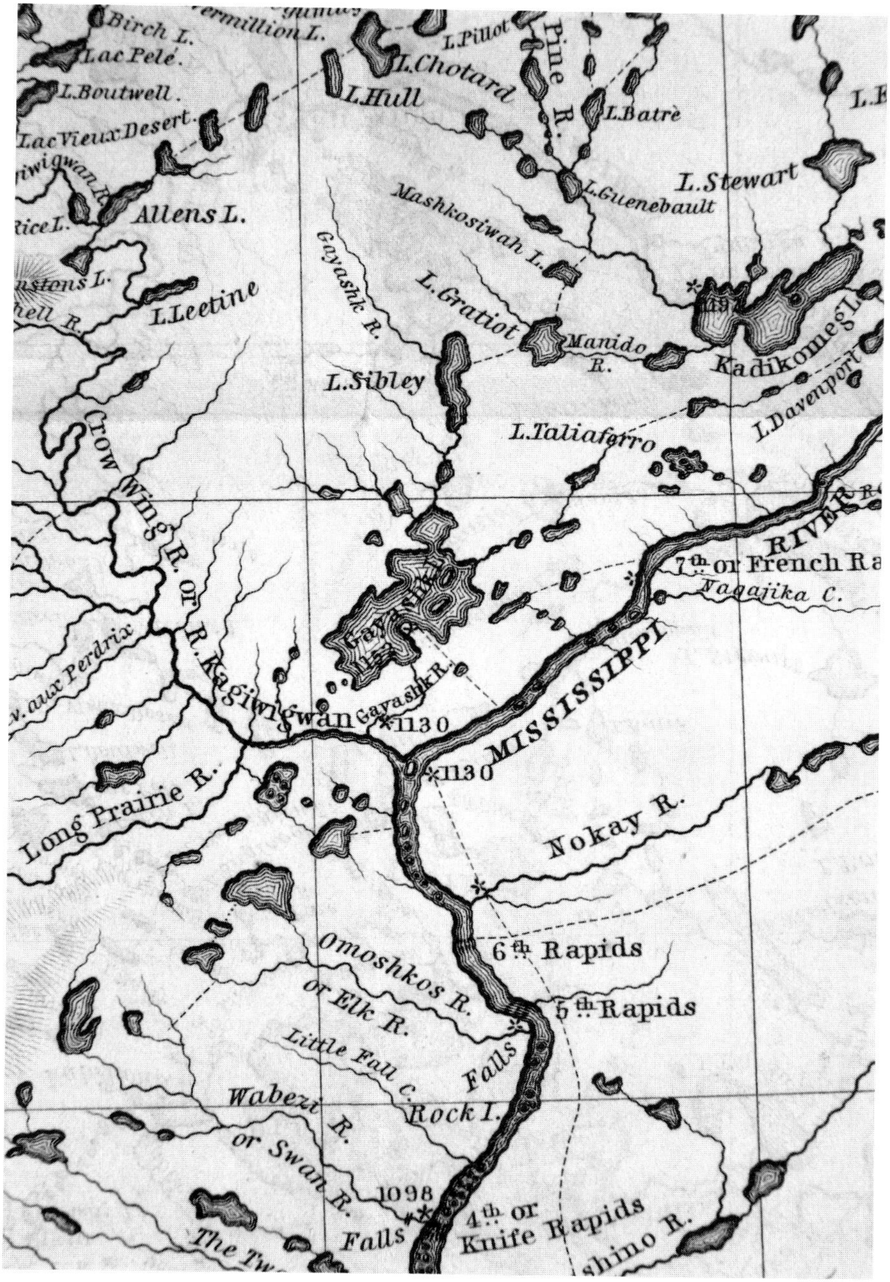

Birch L.
Vermillion L.
Lac Pelé.
L. Boutwell.
Lac Vieux Desert.
L. Pillot
L. Chotard
L. Hull
Pine R.
L. Batre
L. Stewart
Allens L.
Rice L.
L. Guenebault
Mashkostwah L.
Gayashk R.
L. Gratiot
L. Leetine
Manido R.
Kadikomeg L.
L. Sibley
L. Davenport
L. Taliaferro
Crow Wing R. or R. Kagiwigwan
R. aux Perdrix
Gayashk L.
Gayashk R.
1130
RIVER
7th or French Ra
Nagajika C.
MISSISSIPPI
Long Prairie R.
Nokay R.
6th Rapids
Omoshkos R. or Elk R.
5th Rapids
Little Fall C.
Falls
Rock I.
Wabezi R. or Swan R.
1098
4th or Knife Rapids
Falls
The Tw
shino R.

more like a standard geographical work noting topography and geology. There are extensive tables of data and a catalogue of plants. Nicollet was fixing the territory in a scientific grid of sample collections and measurements of altitude, latitude and longitude. Careful measurement was the hallmark of Nicollet's work. He wrote in the *Report*, 'In geographical explorations we can neglect nothing or we are pursued and punished by bitter regret . . . These regrets revive with even more intensity at the period of constructing a map.'[7]

Even in this brief narrative we get some sense of the importance of Native American informants:

> We pitched our tents upon the same prairie, and I had the opportunity to enter into a long conversation with the chief of the party (the Eagle) one of the most intelligent and brave Indians with whom I ever became acquainted. He gave me some very important information in reference to the conduct of our expedition across the prairies, so as to avoid any unpleasant encounter with the warlike parties that meet here during the hunting season.[8]

Fremont's expeditions

Under Nicollet's tutelage, Fremont received solid scientific training as well as invaluable experience of expeditionary life. After their journeys, they returned to Washington to work on drafting the *Map* and the *Report*. In the nation's new capital Fremont encountered the third influential figure in his life, Thomas Hart Benton, a powerful Missouri senator and strong advocate for the Republic's westward expansion. Benton had three daughters; one of them, Jessie Ann Benton, was his confidante and secretary. She was well schooled in Washington politics, ambitious and a polished writer. Jessie Benton married Fremont in 1841. Fremont was now part of a powerful Washington family; his father-in-law was an influential senator and his wife an accomplished woman who wrote most of his reports and much of his memoirs and who worked hard to promote his career. Placed by Poinsett, schooled by Nicollet and now allied to Washington society, Fremont was poised for take off.

In 1842 the Corps launched an expedition into the Rockies to survey a route used by settlers. Benton promoted the idea. Fremont was put in charge. In St Louis Fremont hired Kit Carson and 21 'voyageurs' (men who travelled through the country on behalf of the fur companies). They went west to South Pass and then continued on to climb and name Fremont Peak in the Wind River Ranges. Illustration 29 shows the route followed by the expedition, marked on the map as *F. R. 1842*. After six months of travel Fremont returned to St Louis on 17 October. It was not a path-breaking journey: Fremont travelled on well-used routes; his narrative account is full of meetings with wagon train emigrants on the same trail; it was more of a mapping exercise than an exploration of virgin territory.

Fremont's report was written at Senator Benton's suggestion in the form of a guide, accompanied by a series of maps listing practical information for emigrants. Jessie Benton wrote it, and it was delivered to Abert in March 1843. The German cartographer Charles Preuss illustrated it with pictures of plant specimens as well as with meticulous maps. Written in the form of a narrative diary, the exposition benefited from Benton's good writing skills, possessing an energy and life often lacking in official reports. It soon became very popular. The lively account details a series of cartographic encounters. Soon after the expedition began its travels, on the evening of 28 June, three Cheyenne came to their camp.

> After supper we sat down on the grass, and I placed a sheet of paper between us, on which they traced rudely, but with a certain degree of relative truth, the watercourses of the country, which lay between their villages, and us, and of which I desired to have some information.[9]

We also get more indirect accounts. On the morning of 22 July an unnamed Indian guide refused to lead them any further unless he saw the horse Fremont had promised him. The guide was making sure that Fremont was going to honour a bargain, an indication of bargains unfulfilled in previous cartographic encounters. The guide had some power. Fremont noted, in a sentence that unwittingly reveals who was the real pathfinder, 'I felt strongly tempted to drive him out of the camp, but his presence appeared to give confidence to the men.'[10]

After his successful expedition Fremont was soon back on the trail. In March 1843 Abert issued orders to Fremont to find another pass in the Rockies, south of the area that he had crossed previously, and to explore the territory as far west to the south of the Columbia River. The survey was part of the US probing of the boundaries with Mexico and Britain, opening moves in the imperial drive to gain control of all the territory from 'sea to shining sea'. Fremont was to complete survey work of the region that had first been attempted the year before in the Wilkes Survey. So in 1843 Fremont was again in St Louis recruiting men for the western explorations. He hired many from the previous year, including Kit Carson and Charles Preuss. Over the course of the next fourteen months he travelled through the Rockies, passing close to the Great Salt Lake, up to the Columbia River, south through California before heading back east. On his return to St Louis in August 1844 he was treated to a hero's welcome. Again his travels were turned into a report and map that increased his recognition and national popularity. From Fremont's field notes, Jessie again wrote a driving narrative, which was presented in March 1845, and the Senate ordered 5,000 copies of a volume that combined the reports of the first as well as the second expedition. Two newspapers

29 Detail from Charles Preuss's *Map of Oregon and Upper California . . .* (Washington, DC, 1848).

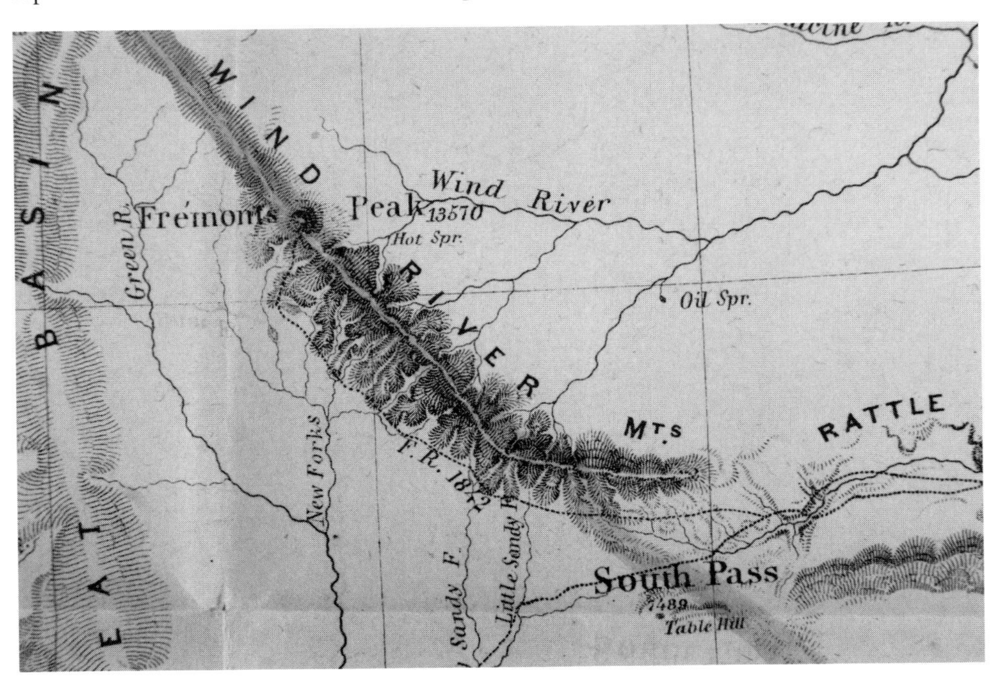

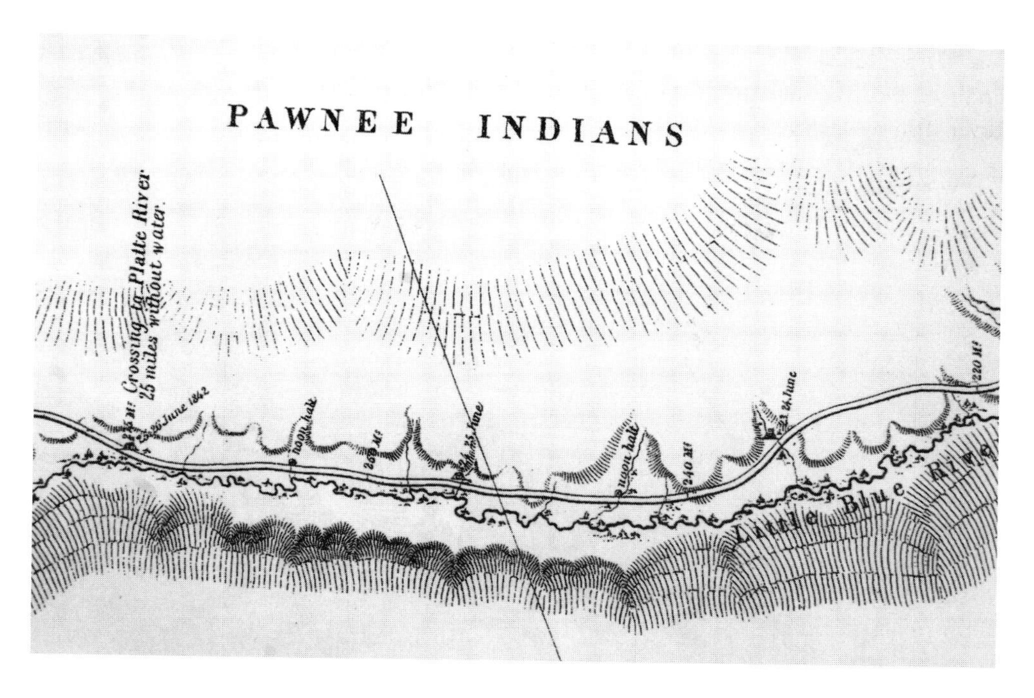

30 Detail from Charles Preuss's *Topographical Map of The Road from Missouri to Oregon* . . . (Washington, DC, 1846).

printed excerpts as well as fulsome reviews. The *Map* and *Report* appeared in 1846 as a book that quickly became a bestseller. Fremont's reputation was now established. He was seen as the embodiment of the heroic dynamic nation pushing ever westward. Readings of the report encouraged the Mormons to move to what would become Salt Lake City and drew attention to the weakness of the Mexicans in California. Preuss's seven-sheet map of the Oregon Trail, drawn in part from Fremont's field reports and published in 1846, gave invaluable advice to migrants making some or all of the 2000-mile journey. Illustration 30 shows some of this trail with Fremont's camps clearly marked.

Fremont outlined the promise of a bigger Republic and promoted a greater imperial expansion. In his wake more settlers moved west and more Washington politicians conceived of a Manifest Destiny made real. Reading his reports today we can see the cartographic encounters that undergirded his explorations. Throughout his journeys he was dependent on Native American guides, Native American paths, Native American directions and advice. On 16 January 1844 he recorded,

> They [Native Americans] made on the ground a drawing of the river which they represented as issuing from another lake in the mountain three or four days distant, in a direction a little west of south; beyond which they drew a mountain; and further still two rivers; on one of which they told us that people like ourselves traveled.[11]

Later that same month, on the 29th, Fremont records a long discussion with Native Americans about how to get through the mountains. They told him he had to find a pass further to the south in order to make a midwinter crossing. It is obvious from a careful reading of the *Report* that, whenever the topography became complex, Fremont was totally dependent on Native American informants and guides.

This second expedition marks the high point of Fremont's career. The reports and maps were widely circulated, and he achieved national recognition. With the benefit of hindsight his later life intones a coda that disappoints in relation to the unalloyed success of his earlier exploits.

Fremont began another western expedition in 1845. He received orders to survey the area around the Red River. Almost 60 men joined Fremont. The party also included Lt James William Abert, son of the Corps' chief. The third expedition travelled to California, north to Oregon and then returned to California where Fremont joined in the American rebellion against Mexican rule in the so-called Bear Flag rebellion. He disobeyed a commanding officer, S. W. Kearney, and was taken back to Washington, where he was found guilty of mutiny. He resigned from the army, led two more privately financed railway survey expeditions to the West, one in 1848 that ended in disaster and another in 1853 that was a useful platform to mount an unsuccessful presidential campaign in 1856. He returned to the army at the start of the Civil War, but did not have great success. He resigned, toyed with the idea of another presidential campaign in 1864, but stepped aside in favour of Lincoln.

It is at the very beginning of the third expedition that we come across Tah-Kai-Buhl – but not in any of Fremont's writings. His court martial meant that he and Benton did not write an official report as they had with the two previous expeditions. The whole journey is discussed only briefly in his *Memoirs*, published in 1887, long after the events had passed into distant memory. He describes trying to start the expedition from Bent's Fort in August 1845: 'I had endeavored to obtain the services of an Indian who knew well the country, and was a man of great influence, especially among the Camanches, but no offer that I could make him would induce him to go.'[12]

We have another witness: young James William Abert, son of the Corps' leader, who had been assigned to the Corps in 1843. He undertook an extensive survey of the northern lakes in 1834 and 1834 before being assigned to serve in Fremont's third expedition. At the start of the expedition, in 1845 at Bent's Fort, Fremont decided to split his party in two, putting Abert in charge of a group that was to survey and map the Canadian River to its source while Fremont went on the California. Abert followed orders, following the Canadian River as far as its confluence with the Arkansas River. The party never met as Fremont pushed on to California and Abert returned to Washington. Abert's official report of his travels, entitled *Through the Country of the Comanche Indians in the Fall of the Year 1845*, was first published in 1846.[13] From this report we read that,

> I endeavoured to obtain information with reference to our southern route, but the only person who knew much of the country through which we should have to pass was a Kioway Indian, called in his own language Tah-kai-buhl . . . Captain Fremont made him most tempting offers to undertake the guidance of our party through these hostile Indians but he refused to go with us.[14]

Remember this is at the very start of the expedition. They are not stuck in the middle of some unknown mountains or sun-scorched desert. They are at a fort, at the beginning of their journey and yet barely able to move, it seems, without the help of Tah-Kai-Buhl. Abert's report, less concerned with self-promotion and career advancement than Fremont's reports, provides a picture of the under-reported role of Native Americans and the heavy reliance on them as informants and guides.

On 8 September Abert notes 'The Indians are remarkable for the skill displayed in their selection of their trails which are always the most practicable routes through the country.'[15] On 18 September at an Indian village at Antelope Butte he records,

> We now placed before them a map of the country, which had been made out at Bent's Fort by Tah-Kai-Buhl . . . Quite a council was called to decide whether or not Buffalo Creek runs into Red River or into the 'Goo-al-pal' as represented on the map. A clean sheet of paper was produced, and the map drawn according to their directions. Tah-Kai-Buhl's map was corrected and the relative portions of the various topographical features were preserved in a surprisingly exact manner, when we consider that it covers an area of about 800 miles in length.[16]

Fremont has left a mark on history because of his and his wife's writings as well as of the writings of others. He is portrayed as an active agent in the making of history. Tah-Kai-Buhl, in contrast, remains a marginal figure, unknown even to most scholars of the US West, seen always the from the perspective of the dominant culture, his image drawn by a US Army officer. His name appears again in *Blackwood's Edinburgh Magazine* in 1848 in an article entitled 'Life in The Far West' by George Frederick Ruxton, an Englishman who was stationed in Canada before he resigned his commission to travel in Mexico.[17] A quixotic figure who later journeyed to Africa, Ruxton described his travels for the magazine. During his visit to Bent's Fort Ruxton came across Tah-Kai-Buhl, whose name according to Ruxton means 'he who jumps'. Again it is a passing reference yet indicates that Tah-Kai-Buhl was an important figure in the encounters between native and non-native peoples. From 1833 to 1849 Bent's Fort was an important hub on the Sante Fe Trail.[18] Established by the Bent St Vrain trading company it was a place where trappers, the army, travellers and indigenous people could trade and sell. Beaver pelts and buffalo robes were exchanged for blankets and cloth, bells and beads, tools and gunpowder. Information and geographical knowledge was also an important commodity for those heading west. As a gifted local expert Tah-Kai-Buhl was often the first point of reference for many trying to make their way through the unknown territory, including Fremont the Pathfinder.

While the Fremonts are recorded and discussed, the Tah-Kai-Buhls continue in obscurity. But as a careful reading of the reports reveal, the exploration and mapping of the West was a cartographic encounter totally dependent on the Tah-Kai-Buhls. We have largely forgotten the Native American side of the cartographic collaboration. Central to all explorations in a region Tah-Kai-Buhl yet remains at the margins of historical accounts. And all the other Tah-Kai-Buhls have yet to receive the due recognition of their pivotal role in the exploration and mapping of the West.

Yakima Pass
Mt. Rainier
Lieut. R. Arnold 1853
G. Gibbs 1853
Atahnam R.
Ft. Simcoe
Yakima R.
St. Helens
Mt. Adams
YAKEMA INDIANS
Gov. Stevens 1853
Dalles
WASCO IND.s
Mount Hood
Fall River
WALLAH-WAL
SCATTERED BANDS
OF INDIANS
Tysch Prairie
SHOSHONES
INDIANS

8 | 'Warren's Map'

In 1854 at the young age of 24, Gouverneur Kemble Warren, an officer in the Corps of Topographical Engineers, was given the difficult job of preparing a general map of the region west of the Mississippi. Warren was instructed to read all the reports and study all the maps produced by the Corps as well as other relevant surveys and expedition reports. Warren drew upon this vast amount of material as well as his own personal experience of survey work. He made three trips to the Nebraska territory and was chief topographic officer in an army expedition against the Sioux.

The map appeared in 1858 as *Map of the Territory of the United States from the Mississippi to the Pacific Ocean*. Warren's map, constructed to the scale of 1:6,000,000, is a major work (illus. 31). A territory that scarcely 50 years earlier was largely unexplored and unknown to science was now brought under one comprehensive piece of detailed cartography. It has been described as 'a great scientific achievement . . . One of the important maps of American history'.[1] Warren's general map is the summation of half a century's worth of exploration. It shows carefully delineated relief and drainage, it represents a land measured, controlled and ripe for development. It was the basis for a commercial map published in St Louis in 1859 by McGowan and Hidt that highlighted routes to the newly discovered goldfields in Colorado.

31 Detail from Lt G. K. Warren's *Map of The Territory of the United States from The Mississippi to the Pacific Ocean* (New York, 1858).

On the bottom right-hand of the map, under the heading in block capital letters 'AUTHORITIES', is a list of 45 sources used in compiling the map. The list is chronological, begins with Lewis and Clark and ends with the current surveys of the Land Office. It includes Long, Nicollet and Fremont as well as the work of Wilkes and the Pacific Railroad Surveys. It is a record of white explorers, surveyors, scientists and military officers. No Native Americans are listed as authorities, yet, as I will show in this chapter, the collaboration with Native Americans underwrote these works that resulted in Warren's cartographic achievement. I will read backwards from Warren's map through some of the published and unpublished reports of the sources that he used in order to uncover the Native American contribution. I will look at expeditions by Charles Wilkes, James Simpson, Randolph Marcy and the Pacific Railroad surveys.

The Wilkes explorations in Oregon

The ninth entry in Warren's list of authorities is 'Com. C. Wilkes, U.S.N., Explorations in Oregon', a naval exploration under the command of Lieutenant Charles Wilkes. Six vessels sailed from Norfolk, Virginia, in 1838 with crew, illustrators, artists, scientists and mapmakers on a trip around the world. It was an act of both grand geopolitical positioning and scientific curiosity that echoed the late eighteenth-century grand explorations of Britain's Captain Cook, the Pacific voyages of France's La Pérouse and D'Entrecasteaux and Spain's Malaspina expeditions.

The US Exploring Expedition covered the globe; they sailed to Madeira, then crossed the Atlantic, rounded the southernmost tip of South America, sailed up the coast to Chile, crossed the Pacific to Tahiti, made their way to Australia and New Zealand, sailed south to Antarctica, turned north to the warmer waters of Hawaii and then east across the Pacific Ocean to Oregon and California. Next they departed from San Francisco, sailing west to Manila in the Philippines through the archipelagos of Southeast Asia, round the Cape of Good Hope and back across the Atlantic, landing in New York in 1842. They returned with thousands of specimens, samples, drawings and maps that formed the basis for many of the national collections now in the Smithsonian. Wilkes wrote a report of the expedition, *Narrative of the United States Exploring Expedition*, published in 1844–5.[2]

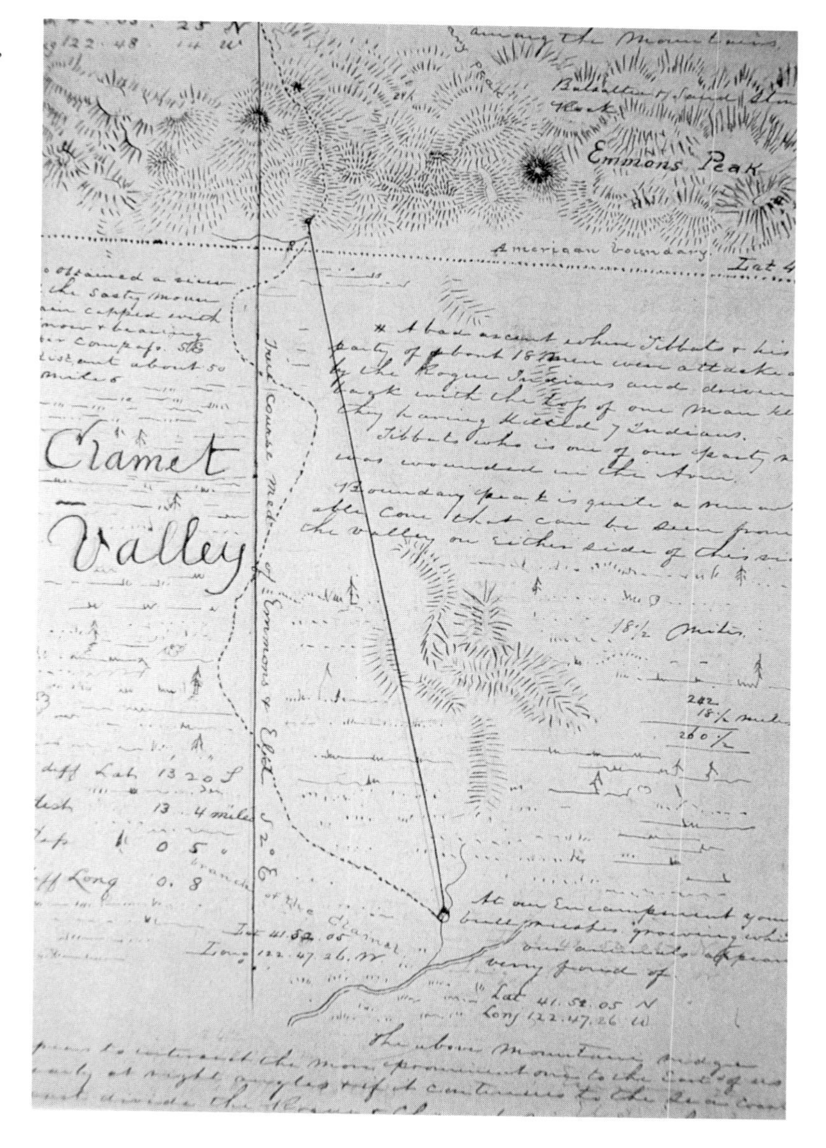

32 Page from Henry Eld's *Journal, Statistics, &c., in Oregon and California* (1841).

In 1841 the squadron was stationed at the mouth of the Columbia River on the Oregon coast. Wilkes drew upon the knowledge of two Chinook men (he called them Ramsey and George) as pilots to sail into the mouth of the Columbia River 'to point out all the localities and channels'.[3] Surveying parties were sent up the river and to the Olympic peninsula. Wilkes also sent a party of men under the command of George Emmons down the Willamette River and then south to what is now San Francisco, where they were to rejoin the squadron. These surveys formed the basis of a number of maps that when knitted with Fremont's map allowed Warren to map the entire region.

Since Wilkes was on board ship most of the time, his official report provides few details of the land surveys. However, we do have another source. Henry Eld, a midshipman on the expedition, was part of the Emmons party that travelled overland to San Francisco. Eld kept a journal of his journey with entries from 6 September to 29 October 1841.[4] It is a linear geography with daily written entries interspersed with figures and maps as well as latitude and longitude measurements and estimations of distance travelled each day (illus. 32). The journey covered approximately 800 miles. The party, which included the scientist James Dwight Dana, naturalist Titian Ramsey Peale, five trappers, an 'Indian' hunter, a 'half breed' as well as a 'Spanish' guide, also had three Native American women, described as squaws, who are all described as capable of bearing arms. Emmons's party was also careful to take an 'Indian' guide named Gurdapi. He was an important figure, his moods carefully monitored. At one stage Eld remarked, 'our guide Gurdapi in particular has been excessively sulky'. Throughout the journey Eld and his companions were dependent on Native Americans as suppliers of food and information. Along the journey they traded for information, food and artefacts such as bows, arrows and elk skin robes in return for axes and hatchets. Eld's journal reveals less the mapping of an empty space or the discovery of a new land, more a journey though a humanized landscape, a peopled land. Reading Eld's more intimate journal we realize that the 'expedition' is more a journey made with Native Americans, guided by Native Americans through Native American territory.

Military mapping exercises

As the frontier moved westward the US military undertook numerous expeditionary surveys into new lands, some purely military, others surveying ventures, most a mixture of both. Many of them added to

33 Illustration of 'Head of Ke che-ah-qui-ho-no, or the main branch of Red River' from Randolph Marcy, *Exploration of the Red River of Louisiana in the Year 1852* (Washington, DC, 1854).

the basis for Warren's map. These included Stansbury's 1849–50 exploration and a survey of the Great Salt Lake, Pope's 1849 expedition to the Red River and the Sitgreaves 1851 expedition down the Zuni and Colorado Rivers. Here I will discuss only two: the Simpson 1849 reconnaissance in the southwest and the Marcy 1852 exploration of the Red River.

James Simpson was a first lieutenant in the Corps of Topographical Engineers assigned to a military expedition sent out in 1849 against the Navajos. A total of 175 men, including four infantry companies and two artillery companies under the command of Lt-Col. Washington, were sent to punish the Navajos for their raids on settlements. Simpson, along with three assistants, was to survey the country 'as the movement of troops will permit'. Simpson documented his adventures in a published account.[5] This account is interesting because it takes place against the background of a punitive expedition whose primary mission was to enforce control and discipline the Navajos. While Simpson was mapping, Washington was leading attacks against the Navajos. The party left Santa Fe on 15 August 1849. They headed west past the pueblo of Jemez and into Chaco Canyon. Six Navajo were killed in a skirmish in the foothills of the Chuska Mountains. The expedition moved further west into the Navajo heartland, where a further skirmish took place. Washington was able to defeat the Navajo and make them submit to US authority. Against this background Simpson, often with military escorts, surveyed the territory, and located many of the abandoned pueblos of the Anasazi culture. The artist Richard Kern drew these in stylized representations.

Simpson's journal is one of the most heavily illustrated reports. It contains illustrations of topographies, geological formation, ruins, hieroglyphics, pottery finds and building diagrams as well as portraits of Native Americans. The brothers Richard and Edward Kern, who had worked with Fremont, were part of Simpson's team and their illustrations did much to convey the abandoned pueblos with a sense of majesty. From the text of the journal it is clear that Simpson used Native American guides. At one point he mentions – but only in passing – that a pueblo Indian, Hosta, explained much of the hieroglyphics to him. His laconic and elliptical descriptions give us only a limited discussion of cartographic encounters, and the report mentions no other guides or informants. In this book I focus on the more explicit articulations of cartographic encounter, but we might want to speculate as to why some writers were so reticent; perhaps they were simply terse writers, or were unwilling to note or simply unable to see the role of Native Americans. The silences may be as revealing as the articulations.

The Simpson journal has to be read in its context rather than in its surface details. In one revealing section the author notes: 'What is of no inconsiderable value, the troops have been able to penetrate into the very heart of their country and thus geographical knowledge has been obtained, which cannot but be of the highest value in any future

34 John Young, 'Cañon of Psuc-See Que Creek Near Camp 41 A', from *Pacific Railroad Reports*, volume VI (Washington, DC, 1857).

military demonstration it may be necessary to make.'[6] His remarks suggest more of geographical conquest than a cartographic encounter, but the result is the same. New and better geographical knowledge enables hegemony to be extended and strengthened.

Captain Randolph Marcy led five expeditions in the West. Although less well known than Fremont, Long or Pike, he is a significant figure in western mapping. In 1852 he led an expedition into the basin of the Red River. From the official report we can easily identify a series of cartographic encounters that aided Marcy in becoming the first white man to trace the river to its source. The expedition was in part driven by scientific curiosity, with geology being a prominent element (see illus. 33). Prior to departure, he wrote in the first paragraph of his report that he could not get a chronometer; so he made observations with his pocket lever watch. No such substitutes were available for local informants.

On 7 May he came across a hunting party of around 150 Wichitas. After telling them of the immense power of the Great Captain (the President) and accompanied by the usual ritual of present-giving, he reports,

> I made inquiries concerning the country through which we still have to pass in our journey. They said we would find one more

> stream of good water about two days travel from here; that we should then leave the mountains and after that find no more fresh water to the sources of the river. The chief represented the river from where it leaves the mountains as flowing over an elevated prairie, totally destitute of water, wood, or grass, and the only substitute for fuel that could be had was the buffalo 'chips'. . . . I inquired of them if there were not holes in the earth where the water remained after rains. They said no, that the soil was of so porous a nature that it soaked up the water as soon as it fell. I then endeavored to hire one of the old men to accompany me as guide; but they said they were afraid to go into the country, as there was no water, and they would be fearful they would perish before they could return.[7]

Only a month later, in early June, did he realize that the cartographic encounter was not completely accurate:

> as we ascend the river we have conclusive evidence of the falsity of the representations of . . . the Wichitas . . . Their statements have proved false in every particular, as we have thus far seen the country well watered, the soil in many places good, everywhere yielding an abundance of the most nutritious grasses, with a great sufficiency of wood for all the purposes of the traveller.[8]

The Wichitas gave information to dissuade the military from going there. They wanted to suppress, not encourage, further white settlement. Not all cartographic encounters were based on the exchange of honest information.

Despite the deviousness of the Wichitas, Marcy continued to use Native American guides and informants. On 1 June he writes, 'Taking an old Comanche trail this morning, I followed it to a narrow defile in the mountains, which led me up through a very tortuous and rocky gorge, where the well-worn path indicated that it had been travelled for many years.'[9] Marcy waxes eloquently about one guide, a Delaware, possibly John Bushman:

> the very moment he takes a glance over a district of country he has never seen before, he will almost invariably point out the

> particular localities (if there are any such) where water can be found, when to others there seems to be nothing to indicate it. Such qualifications render the services of these people highly important, and almost indispensable in a tour upon the prairies.[10]

Marcy's report contains two maps: one of the country between the frontier of Arkansas and New Mexico and one of the Red River basin. These maps informed the great Warren map. Marcy also played an even larger role in the westward expansion. His 1859 book *The Prairie Traveler*, with the informative if long subtitle, *Handbook for Overland Expeditions, with Maps, Illustrations and Itineraries of the Principal Routes between the Mississippi and the Pacific*, became a well-known and well-used text guiding emigrants safely westwards and remained popular through the rest of the century. An internet search finds ample evidence of Marcy's exploits, but almost nothing about the work of John Bushman.

The Pacific Railroad surveys

Warren drew heavily upon the surveys undertaken to find a railway route to the Pacific. Finding such a route across the high plains and through the Rockies was a contentious issue. A railway would increase land values in areas close to the line and depress them in areas far from the railway's final route. Powerful economic, financial and political interests sought to influence the decision. To break the deadlock Congress agreed on a scheme for multiple routes to be surveyed by the military with scientific help. Four routes were surveyed, and the reports were published between 1855 and 1860 in the multi-volume *Pacific Railroad Reports* (*PRR*). The *PRR* is one of the great documents of western exploration and includes copies of orders, official letters, all the reports and a vast array of maps, drawings, pictures, profiles (illus. 34–38). It is one of the largest scientific inventories and artistic depictions of the American West, providing a visual sense of the new lands for the public and politicians back East.[11]

Isaac I. Stevens was responsible for the northern survey, from St Paul, Minnesota, to Puget Sound, a distance of approximately 1,864 miles that lay between the 47th and 49th parallels. Stevens was a West Point graduate and served in the Corps of Engineers during the Mexican–American War. He was made governor of Washington

Territory in 1853. The same year he was also put in charge of the northern survey. The official report appeared in volume 1 of the *PRR*. As the survey team moved westward they came to the complex and often bewildering topography of the Rockies. Stevens ordered one of his young lieutenants, John Mullan, to go from Fort Benton to the camp of the Flathead Indians and obtain guides to find the best passes through the mountains. The official report contains a detailed journal account by Lt John Mullan. When he left Fort Benton, Mullan took a guide, a Blackfoot known as White Brave, two unnamed Blackfoot Indians and three voyageurs. When the guide took him to the camp, Mullan asked the chief for more guides. The chief gave him five; one turned back, but the rest went with Mullan. It was fall, and the weather was deteriorating. At the very start of the journey, with winter fast approaching in the higher altitudes, the Blackfoot guide, far from home and moving further away each day, was leery of the whole enterprise. In an entry for 18 September 1853 Mullan noted,

35 John Young, 'Abies Douglassii' [Douglas Fir]', from *Pacific Railroad Reports*, volume VI (Washington, DC, 1857).

36 Plate of 'Fishes', from J. H. Richard, *Pacific Railroad Reports*, volume XI (Washington, DC, 1859).

Resuming our journey this morning, I noticed that our guide showed an evident disposition of unwillingness to accompany me further. Through the interpreter he has asked me to release him from his engagement and to allow him to return home. This I refused to do . . . he appeared very sullen and promised to accompany us to the end of the journey. When everything was ready, I told him to mount his horse and come on; he said he wished to smoke and that he would overtake us in a short time. Presuming that he had fully made up his mind to accompany us, I thought nothing of it, but rode on without him; we have not seen him since.[12]

Cartographic encounters could be and often were terminated. I focus on the more expansive but there were many abridged encounters.

Mullan was now more reliant on the Flathead guides, whom he praised. 'I cannot say too much in favor of these men who were with us; they were pious, aged, firm, upright and reliable men . . . they all knew the country well and made excellent guides and good hunters.'[13] By the time he arrived in Fort Owen on 10 January Mullan had relied upon the guides and utilized Native American paths. On 4 January, for example, he remarks, 'We crossed during the afternoon the trail leading toward the Little Blackfoot fork, which is the main Flathead trail across the mountains.'[14] In what now appears an inevitably ironic twist, this route from Fort Benton to the Columbia River quickly became known as the Mullan Route.

A second survey followed a line west between the 37th and 39th parallel from Westport to San Francisco, a distance of 2,080 miles; it was led by Lt John Gunnison until he was killed by the Utes in Utah. Lt Edward Beckwith then took over. Warren was part of this survey team. Beckwith's report, published in volume 2 of the *PRR*, contains little or no mention of cartographic encounters. However we get fleeting glimpses of a guide, Massalino, who led them through the mountain passes, of attempts to procure guides with 'trinkets, cloths, paints and blankets', as well as the use of a Utah Indian guide named Tewip Narrienta or Powerful Earth. And Beckwith also writes, 'heavy Indian trails attest the use they make of this pass . . . it affords an excellent wagon and railroad route . . . as a testimonial to the memory of the officer who explored it [Gunnison] I have given his name to this pass'.[15] A pass 'heavy' with Native Indian trails and traffic, discovered and explored centuries before and used for generations is now turned into a memorial for a fallen army officer. By such acts of toponymic imperialism a Native American landscape was turned into an American national territory.

The most southerly route followed along the 32nd parallel. It was surveyed from San Diego to Fort Washita by two groups: Lt John Parke was in charge of the survey from San Diego to the Rio Grande and Captain John Pope led a team from the Rio Grande to Fort Washita, in present day Oklahoma. Pope's report, published in volume 2 of *PRR*, is very concise. 'I have more or less carefully avoided embarrassing the subject with a narrative of the daily incidents which must be more or less irrelevant.'[16] We are given few opportunities to hear of Native Americans. The report is more a geopolitical one, noting the best routes for railways and the optimum sites for military forts. Not all reports, thankfully, were as laconic as Pope's.

The survey along the 35th parallel from Fort Smith, on the border of present-day Arkansas and Oklahoma, to Los Angeles was led by Lt Amiel Whipple. His official report is printed in volume 3 of the *PRR*. Whipple is a more expansive writer than Pope and casts a wide view over the expedition. The survey team left the fort on 14 July 1853, but by 5 August their guide had to leave because his child was taken ill. Their subsequent journey is only made possible by a series of Native American guides and informants. I will quote from Whipple's journal at length in a chronological order since it provides such a detailed account of the use of Native American guides.[17]

> *August 6*
> Finding a Shawnee Indian at Perry's store we induce him to accommodate us as far as the first Shawnee village, about 25 miles distant.
>
> *August 11*
> We inquired of an Indian for the right road, but gained little satisfaction . . . At length he intimated that we might as well follow him.
>
> *August 12*
> Perceiving that great advantages might be derived from information such as a guide should posses; before leaving Fort Smith

37 Plate of 'Mammals', from J. H. Richard, *Pacific Railroad Reports*, volume XI (Washington, DC, 1859).

38 Plate from 'Birds', from J. H. Richard, *Pacific Railroad Reports*, volume XI (Washington, DC, 1859).

a messenger was sent to engage Black Beaver, A Delaware chief, who however, declined the service . . . I rode to Shawneetown in search of a guide to accompany an exploration . . . but none would be willing to make the trip.

August 13
This afternoon an Indian came into camp who claimed to be a nephew of Black Beaver. He professes to be well acquainted with the country south of the Pawnee hills, and says that to-morrow he will conduct us by a smooth and direct route from our camp.

August 15
At his suggestion, instead of following the train we kept the dividing ridge between the Canadian and the headwaters of the river Boggy.

By the middle of August they need more guides because Johnson, their Shawnee guide, refuses to continue the trip. They send for John

Bushman, 'A Delaware guide of some celebrity' (the same Bushman whom Marcy lauded), and Jesse Chisholm, a Cherokee. Negotiations do not go well. They break down completely with Bushman, who curses them: 'Maybe you find no water; maybe you all die.' Chisholm also declines the service because, as Whipple notes, as a man of considerable wealth, Chisholm is unimpressed with the miserly government rate offered for his services. The remainder of the expedition reads like a comedy of errors as they endeavour to get guides to show them the route:

> *August 24*
> As we were preparing to pursue our journey, another Indian rode into the camp . . . we inquired the direction to the old wagon trail, which being very obscure, we had unintentionally departed from . . . While reconnoitring our position, the Indian guide deserted us. We were proposing to retrace our steps, when another Indian, supposed to be a true Kichai, made his appearance. For a consideration, he showed us the desired route, which was upon the northern slope of the ridge intersecting the branches of the Walnut creek.

Later, in November, they persuade a Zuni chief to provide guides. Throughout February and March they have to rely on the help of Mojave chiefs. On 12 March a complicated mapping exchange takes place:

> In making inquiries of the Indians we were cautious not to incline their minds to any preconceived notions of our own by asking leading questions. But we traced in the sand a depression to denote the valley of the Colorado, and in the middle represented the meandering river which they recognized. Heaps of sand piled on each side, and called sierras, they understood to denote mountains. Our trail was then marked out, indicating the camps and the springs, and the mountains crossed from the Mojave villages to the flowing water of Rio Mojave. Thence we represented the valley of that river by a channel scooped from the sand, indicating where there was water, and where there was none. Then we inquired of the Indians whether or not the channel was lost in the

lake. The guide understood the question, and instantly with his hand cut a passage through the heaps of sand that had been piled upon the ground-map for mountains around the lake, representing a continuous valley to the arroyo by which we haft the Colorado. Pointing along the line traced, he said it was a smooth arroyo, containing no water upon the surface, nor obstructed by hills, but the valley was wide, level and sandy throughout. When asked if the distance was great, he replied that it was; and counting with his fingers, he indicated ten as the number of days required to travel it with our wagons.

'Warren's map'

We can now see the need for quotation marks around 'Warren's map'. Not because of the range of authorities that he cites but because of the Native American contributions to the authorities. The work of the authorities was predicated upon a vast network of Native American trails, life-saving trade with Native Americans for much-needed food and information, and numerous Native American guides who guided the survey teams across the plains, through the desert and over the mountains.

While working on the map Warren was also involved in many expeditions along the western frontier. From 1855 to 1857 he made three expeditions to the Nebraska territory. In 1855 he was the topographical engineer on General Hanley's expedition against the Sioux in present-day Nebraska and South Dakota. He wrote the main report. The next year he led a survey expedition along the Missouri River up to Yellowstone. And the year after that he was on an expedition to the Black Hills.

Warren's official reports reveal the geopolitical realities that underlay the cartographic encounters and allow us to see the symbiotic destruction that we discussed at the beginning of the book. In his report of the 1855 expedition Warren notes, 'Military occupation is essential to the safety of the whites and the military posts should be in such positions and occupied by such number as effectually to overawe the ambitious and the turbulent.'[18] This is perhaps one the earliest references in the US army to what later became known as 'shock and awe'. Warren went on to recommend the construction of an infantry post at

Fort Pierre, well connected to Fort Laramie, in order to drive the Dakotas south. In a letter written in 1858 to George Jones, Secretary of War, Warren argued that in the Black Hills area, 'The wealth of that country is not properly valued, and the Indian title not being extinguished, there is no opportunity to settle it . . . Extinguish the Indian title and extend the protection of the territorial government.'[19]

When Warren was in the Black Hills, the Sioux complained to him of white encroachment into their lands, the negative impact on the buffalo and the need for the territory to be held as Indian land. They had given up so much, they pleaded, that the Black Hills must be left to them. One Native American guided him through part of the territory; 'In return for this he wished me to say to the President and to the white people that they could not be allowed to come into the country. All they asked of the white people was, to be left to themselves and let alone.'[20]

At this late stage in the struggle, a cartographic encounter was proposed with the knowledge that the tide had turned. At the beginning of the seventeenth century in the East, John Smith desperately needed the information of the Native Americans of the Chesapeake. By the mid-nineteenth century in the far West, both explorers and Native Americans knew that the power had shifted. Now the whites, having gained so much information, no longer needed the Native Americans. The era of the cartographic encounter was almost over.

9 | Closing the Frontier in the West

Mapping surveys of the West continued after the Warren map. After the Civil War Congress funded four large-scale surveys: they are commonly known by their principal leaders: Clarence King, Ferdinand Hayden, Wesley Powell and George Wheeler.

The King Survey (1867–73) mapped 15,000 square miles along the 40th parallel. This was an army undertaking but was led and staffed by civilian scientists. The Hayden Survey (1873–6) mapped a large area around Yellowstone. The Powell Survey (1869–78) consisted of a 1,500-mile strip down the Colorado River and a second survey of the Grand Canyon. The Wheeler survey (1871–9) covered almost 350,000 square miles, a huge area west of the 100th meridian. While they differ in some ways, and there was often intense rivalry between the leaders for recognition, scientific credibility and government funding, they share a number of characteristics. First, the surveys mark a shift from the military to civilians. Three of them were largely civilian affairs, while the Wheeler Survey, although initially a military-led undertaking, was merged into the Geological Survey. In part this reflects a more marked division of labour between military officers and scientists and the rise of organized science as a largely civilian enterprise. The scientific and military spheres were getting more specialized and distant. There were fewer like Abert, father and son, who were both scientists and officers. But the shift from military mapping to civilian scientific triangulation also marks a change in the West. Soldiers were needed less because the Native Americans had been defeated and major military incursions were no longer required. With the work of cartographic encounters almost complete, the Native Americans were becoming superfluous. We can read the report of the big four surveys and still find instances of reliance on Native American guides and informants, but less so as time moves on and the more geographical information is collected, codified and represented. As the nineteenth century comes to a close, the Native Americans are represented by two dominant discourses. The first is of a sense that they are passing into obscurity. Powell, for example, developed an interest in the languages of the Native Americans, compiling over 200 vocabularies. He was appointed Indian Commissioner in 1873. No longer a threat, the Native Americans now become a source of ethnological interest, all the more noteworthy because they appear tantalizingly close to the edge of extinction. The United States established the Bureau of Ethnology in 1879. The second discourse is the annoyance with their continuing acts of resistance, futile but still troubling. The Wheeler Survey embodies these competing tensions.

George Montague Wheeler was born in October 1842 in Massachusetts. He graduated from West Point in 1866 and was appointed second lieutenant in the Corps of Engineers and employed in surveying duty in California. He was promoted to first lieutenant in 1867. Four years later he was given charge of the military survey of territory west of the 100th meridian. The survey was to be the great work of his life, absorbing all his energies. Fieldwork took place every summer from 1871 to 1879 and involved fourteen trips in all. A vast area was mapped, surveyed, drawn and photographed. Wheeler drew extensively on Native American guides. Two of them were portrayed by the expedition's photographer Timothy O'Sullivan with Apache Lake in the Sierra Blanca Range in Arizona as the background (illus. 39).

> From among the Utes and Pah-Utes found north and west of the Colorado River, it was possible to obtain friendly guides, many of whom proved most invaluable in pointing out the little hidden springs and streams, especially in the Death Valley country, Southwestern Nevada, and Eastern California sections.[1]

39 Timothy H. O'Sullivan, 'View on Apache Lake, Two Apache scouts in the foreground', 1873, taken on the 'Wheeler Survey'; from Lt G. M. Wheeler, *Photographs Geographical Explorations and Surveys West of the 100th Meridian . . .* (Washington, DC, 1874).

3

40 J. K. Hillers, 'Zion National Park', *c.* 1872–3, taken on the 'Powell Survey'.

But coming at the last stages of mapping and the virtual completion of US territorial hegemony, Wheeler's writing also betrays his frustration with the continuing resistance, tinged with a sense of its futility. On 4 January 1872 he had to write a letter to a David Loring, whose son was killed by the 'murderous hand of the well known Apache who committed this outrage'.[2] Wheeler's personal letters, rather than his official reports, are full of anger at the danger posed by 'Indians'. Wheeler still needed Native American guides but he never trusted them. He describes two Apaches of the Coyotero tribe, perhaps the two men in O'Sullivan's photograph: 'They are true savages . . . as guides these Indians are frequently used by marching troops and scouting parties. They are treacherous, also being faithful only out of necessity and when their actions are subject to constant scrutiny.'[3]

Let us take the Apache perspective for a moment. Pressed up against a frontier that was about to suffocate them, the Apache were running out of options. The fullest cartographic encounter would seal their fate. They had to work with the army but resist somewhat, as full incorporation would doom them to extinction. They had to guide, but not too well. On the other side, officers like Wheeler still needed them but were wary of their continuing resistance. Even a suspicious Wheeler could see the end in sight. In one of his reports he noted that 'the fate of the Indians sealed, the interval during which their extermination as a race is to be consummated will doubtless be marked in addition to Indian outbreaks, with still many more murderous ambushes and massacres'.[4]

The four great surveys produced a wealth of material. The work of survey photographers such Jack K. Hillers, Timothy O'Sullivan and William H. Jackson, illustrators such as W. H. Holmes and painters such as Thomas Moran was widely dispersed, especially via the burgeoning mass magazine industry. Images of the West circulated widely, but it was a romanticized West, one largely empty of its original people (illus. 40). The contested territory had become romantic scenery, a process aided by the indigenous people in a series of cartographic encounters that transferred geographical information and ultimately power from those facing east to those moving west.

PART IV | conclusions

Those natives who accompanied the expedition deserved the highest praise; they were intelligent, faithful . . . and extremely useful.
Thomas Mitchell, 1837

41 A detail from John Smith's *Generall Historie of Virginia* (London, 1624).

10 | Cartographic Encounters in Australia

I have shown that the exploration and mapping of the New World was a process more of collaboration between indigenous peoples and newcomers than is commonly represented. I have drawn on the colonial and imperial experience of North America. My basic argument, however, is not restricted to this region or even to the New World. If we look more closely at the exploration narratives of those travelling through Africa, Asia, Australia and the northern polar region, the same story can be told and retold. To bolster this claim I will consider a sample of the 'white' exploration of Australia.

'I was not prepared for the extent of the desert I encountered.'[1] So wrote Charles Sturt, one of the first white explorers into the Australian desert. Few non-Aborigines were prepared. The continent was vast and, inland from a narrow coastal strip, the territory quickly became a huge, bewildering expanse of very dry land (illus. 43). For the Aborigines it was their home, their source of meaning and a site of sustenance. They had been living on this land for up to 60,000 years. For the recently arrived whites, who came to settle only after 1788, Australia was a formidable, unknown environment, a desert.

A captain in the British Army, Sturt arrived in Sydney in 1827. Governor Darling appointed him to an eleven-man expedition charged with following the Macquarie River. From 1828 to 1829 they travelled up the Macquarie to the Darling River. Soon after their

42 The dry interior of Australia.

return Darling sent Sturt out again, this time up the Murrumbidgee River. Underlying these journeys was a belief in a great inland sea. Sturt still believed in the possibility even after his return from the interior, but found no immediate opportunity to test the idea. His patron, Darling, had left the colony. But in 1844 he was appointed head of an expedition that left from Adelaide that same year. Convinced of the inland sea they dragged a boat with them into the dry centre of the continent.

At the very beginning of the expedition, Sturt hired guides, two Aboriginal men, Camboli and Nadbuck. They played a vital role in getting the expedition through the desert. They had sophisticated geographical knowledge.

> Nadbuck now informed me that we should have to cross the Ana-branch and go to the eastward, and that it would be necessary to start by dawn, as we should not reach the Darling before sunset.[2]

> Nadbuck informed me that by going direct to the opposite point where, after coming up again, it [the river] turned to the north, we should cut off many miles, but that it would take a whole day to perform the journey. I determined therefore to follow his advice.[3]

> We were again led by Nadbuck across the country, to avoid the more circuitous route along the river.[4]

Nadbuck, it seems, was the real leader of the expedition. Nadbuck left the party but Sturt used other guides. His narrative mentions Toonda and Munducki.

The guides were also important go-betweens for Sturt with the local tribes. Sturt's narrative is full of these exchanges. One example:

> Mr. Browne, with the assistance of Nadbuck, gathered a good deal of information from the natives . . . They stated that we should not be able to cross the ranges, as they are covered with sharp pointed stones and great rocks, that we should fall on and crush us to death; but that if we did get across then to

> the low country on the other side, the heat would kill us. That we should find neither water or grass or wood to light a fire with. That the native wells were very deep, and that the cattle would be unable to drink out of them; and finally, that the water was salt, and that the natives let down bushes of rushes to soak it up.[5]

And when there were no guides as go-betweens, he used sign language,

> On my seeking to know by sign, to what point the creek would lead us, the old man stretched out his hand considerably to the southward of the east, and spreading his fingers suddenly dropped his hands, as if he desired me to understand that it commenced, as he showed, by numerous little channels uniting into one not very far off.[6]

Sturt used informants so much he gained a hard-won understanding of the strategies, deceptions and exaggerations that informants could and did employ. At one point he was disappointed that he had fed two 'natives' but still could not get anything from about the 'nature of the country before us'. As he openly documents in his narrative, Sturt was guided through his journey into the interior by a succession of 'native' guides and local informants. He travelled on 'native' paths, drank at 'native' waterholes and relied upon 'native' hospitality. He never did find the great inland sea. The eighteen-month journey was turned back in 1846 at what is still called Sturt's Stony Desert.

Thomas Mitchell arrived in Sydney as the Assistant Surveyor-General in 1827. He and Sturt jockeyed for leadership of expeditions. Mitchell referred to Sturt as an 'amateur traveller'. Between 1831 and 1836 Mitchell led four expeditions into the interior and his accounts describe the extensive use of native informers and guides. At one point he used a woman, Turandurey. His account of the 1836 expedition into the Murray Darling Basin, for example, notes, 'In this camp of preparation I was visited by our friends the natives; and one, who called himself John Piper, and spoke English tolerably well, agreed to accompany me as far as I should go provided he was allowed a horse and was clothed, fed etc; all of which I immediately agreed to.'[7]

Piper proved invaluable. He told the thirsty party to refrain from drinking at a waterhole after he noticed it was poisoned by the locals to get fish. He was an important go-between for the whites in their cartographic encounters with local groups. 'He ascertained from one of these natives that after eight of our daily travels the bed of the Lachlan would contain no water and one must go to the right and that in two days travelling we should reach very great water.'[8]

Mitchell's account is full of the use of local guides and native informants. Information on sources of water, and the direction of travel was entirely based on local informants. Mitchell's was less an exploration than a series of encounters that enabled passage across difficult terrain. Mitchell freely admits the importance of John Piper, 'In tracing lost cattle, speaking to the wild natives, hunting or driving, Piper was the most accomplished man in camp . . . and in authority he was allowed to consider himself almost next to me, the better to secure his best exertions.'[9]

Edward John Eyre made a number of expeditions in Australia. In 1839–40 he travelled into the interior from Adelaide, a lake there still bears his name. In 1840–41 he went west across the Nullabor Plain to Albany in western Australia. Eyre's accounts of his journeys are remarkable for the attitudes toward Aborigines that strike the modern reader as more in line with thinking now than with thinking then. He writes of white intrusion and aggression and the dispossession of the land. Eyre is surprisingly frank about black–white relations. His views appear in his 1845 *Journals of Expedition and Discovery*, which is part description of the travels and part discourse on the plight of the Aborigines. Eyre was open about his reliance on the indigenous peoples. In his first expedition he noted,

> During the forenoon we were visited by a party of natives, who came to get water at the hole in the sand. They were not much alarmed, and soon became very friendly, remaining near us all night; from them I learned that there was no water inland, and none along the coast for two days' journey, after which we should come to plenty at a place called by them 'Beelimah Gaip-pe'.[10]

'Guided by the natives, we moved onward through a densely scrubby country.'[11] In early January 1840 he noted a typical encounter,

> In the evening we made many inquiries of the natives as to the nature of the country inland . . . though we were far from able to understand all that they said or to acquire half the information that they wished to convey to us, we still comprehended them sufficiently to gather useful and important particulars.[12]

In 1840 Eyre travelled from Port Lincoln in South Australia to Albany in the southern part of Western Australia. He had to cross the parched Nullarbor Plain. His companions after the initial stage were one white overseer, John Baxter, and three Aboriginal guides, Cootacha, Neramberein and Wylie. All four had travelled with Eyre on his previous journeys. It was a difficult crossing. With supplies dwindling, hope running out and Eyre away from the camp, the Aborigines killed Baxter and left the camp. Eyre followed them and persuaded Wylie to stay with him. Eyre recounts the events with some sympathy, surmising that the 'boys' were thirsty and exhausted and probably tried to leave with the provisions. A struggle probably ensued and that was how Baxter was killed. Eyre writes that most people would have done the same thing. Wylie and Eyre carried on. They arrived in Albany on 7 June 1840.

Ludwig Leichhardt was a German who landed in Australia in 1842. He had a scientific bent and organized an expedition from Moreton Bay in Queensland to Port Essington in the far north. It was partly financed by pastoral interests eager to discover new grazing area. Leichhardt's party left a cattle station on 1 October 1844. The party consisted of two Aborigines, Harry Brown and Charlie Fisher, and seven Englishmen. Leichhardt took his time collecting plants. After only a month two Englishmen had deserted and the initial provisions were running low. It took them over a year to arrive at their destination, the northern coast of Australia, on 17 December 1845. The last months were especially difficult as they were exhausted and starving. Recounting the last month Leichhardt noted the importance of native informants, 'My first object was to find good water and our sable friends [two Aborigine men Eooanbery and Minorelli] guided us with the greatest care pointing out to us the most shady road to some wells surrounded with ferns.'[13] Close to starvation Leichhardt's party was guided and kept alive by the local people. Without the locals the travellers would have died.

And that fate did overcome Leichhardt in his 1848 journey across the continent from east to west.

At his third attempt in 1862 John McDouall Stuart succeeded in becoming the first white man to cross the continent of Australia from south to north. His first attempt in March 1860 ended after heavy rains, illness in the party, Stuart's right eye being severely damaged, and an attack from Aborigines that forced them back. A second attempt, funded in part by the government of South Australia, was abandoned because of dwindling provisions. The successful expedition was another government-financed party that set out in January 1862; they reached the northern coast on 24 July. On his return to Adelaide in January 1863 Stuart was given a hero's welcome.

Few people have the time or inclination to read Stuart's original reports. A more popular solution is to read the contemporary compilation. Tim Flannery's *The Explorers*, published in 1998, provides an Australian sampling of over 60 explorers' reports. The three-page Stuart entry includes a brief encounter with some indigenous people but they figure more as a form of passing curiosity, an exotic part of the journey, unconnected to the movement through space. However, if we read the original account we gain a much deeper sense of cartographic encounters. Here is Stuart reporting an event on 6 May 1862 while in the desert interior: 'Yesterday they were visited by a few natives who seemed to be very friendly . . . They pointed to the west as the place where they got the bamboo and water.'[14] Knowledge of water was vital in this hot dry environment. Two days later, early in the morning, after leaving camp, Stuart noted that

> About a mile beyond, struck a native track, followed it, running nearly north-west, until nearly three o'clock p.m., when we came upon a small water hole or opening in the middle of small plain. Which seemed to have been dug by the natives, and is now full of rain water.[15]

The party is directed toward much needed water by Aborigines who 'point to the south-south east and make signs by digging with a scoop that there is water in that direction'. Stuart reports numerous cartographic encounters, 'Tried to make them understand by sign that I wanted to get across the river; they made signs by pointing down the river, by placing both hands together having their fingers closed . . . which led me to think that I could get across further down.'[16] Trading fishhooks for information Stuart made his way across the continent through these cartographic encounters.

As the nineteenth century progressed white explorers advanced further into the desert interior. Here water was a vital yet scarce commodity. One of the explorers, Ernest Giles, who travelled through the very dry central and western deserts, summed it up: 'I should greatly like to catch a native; I'd walk him off alongside my horse, until he took me to water.'[17] Giles was the more typical nineteenth-century explorer and narrator; 'natives' were a problem to be overcome, or resource to be exploited; no sensitive analysis as with Eyre. However, Giles's narrative also reveals his reliance on their knowledge of water: 'Found a place where the natives had recently dug for water . . . so I decided to encamp there for the night greatly pleased with having been so fortunate in our day's travel.'[18] He followed 'natives' to water. At one point, when his party approached a waterhole, one native called to him 'Walk whitefella, walk.' Giles was drinking up a vital previous resource. But his response was to take his rifle and fire at random. On Christmas Day 1873 a group of Aborigines shouted abuse at them. Giles at least had the self-awareness to understand their rage, 'He most undoubtedly stigmatized us as vile and useless . . . took upon ourselves the right to occupy any country or waters we might chance to see . . . killed and ate wallabies thereby depriving him and his friends of their natural and lawful game.'[19]

The Aborigines were informants, guides, linguists, translators, geographers, go-betweens, providers of water and food. The white exploration of Australia was predicated upon a heavy reliance on the local people. Mitchell summed it up best, 'Those natives who accompanied the expedition deserved the highest praise; they were intelligent, faithful . . . and extremely useful.'[20] However, while the names of Sturt, Eyre, Stuart, Giles and Leichhardt are remembered and often inscribed in the landscape, those of Munducki, Nadbuck, Toonda, Turandurey and John Piper are now long forgotten or rarely recognized.

11 | Journey's End

In this book I develop the notion of cartographic encounters. I argue that the exploration and mapping of the New World, and indeed most of the world, was a process more of collaboration between indigenes and newcomers than is commonly represented. The image of the lone, western hero exploring a virgin territory is so far off the mark as to be laughable. But remnants of this image persist. In reality, exploration was a movement through a populated, inhabited land and the newcomers drew upon local knowledge, used existing trails, employed local guides and relied on local informants. It was less an expedition through a wilderness than a journey through a peopled landscape. The journey was dependent on the guidance, knowledge and help of local people. Explorers were more negotiators than pathfinders. The paths were already there. Indigenous people acted as guides, cartographers, pathmakers, suppliers of much-needed food and geographical information: they were an integral part of the exploration and mapping of the New World.

43 'Itasca Lake', *c.* 1851, from Henry Schoolcraft, *History of the Indian Tribes of the United States* . . . (Philadelphia, 1857).

Despite the dominant view that Europeans created maps of the continent on their own, Native Americans helped considerably in the mapping of North America. Rather than a simple cartographic appropriation by Europeans, it is more accurate to consider the notion of cartographic encounters involving Europeans and Native Americans. The mapping of the continent was underpinned by native knowledge. All the explorers' accounts tell the same story: the Europeans landed in a populated place; to find their way and move around the unfamiliar territory they relied upon the indigenous people to provide information, advice and guidance. In return the indigenous people obtained trade goods, spatial information and the possibility of new alliances against long-standing enemies. This exchange was invaluable to the Europeans over both the short and long term. For the Native Americans, in contrast, the exchange over the short and medium term could prove useful but over the long term it sealed their fate.

The cartographic encounters resulted in Native Americans losing control over their territory. No longer needed, they were moved and marginalized. And to add insult to injury, the role of Native Americans in the exploration and mapping is still too often ignored or forgotten. The dominant discourse is one of Fremont's exploits rather than the knowledge of Tah-Kai-Buhl, the maps of William Clark rather than those of Ac ko mok ki while the landscape is inscribed with the name of John Mullan rather than of White Brave.

There has been a conscious removal of the Native American contribution. While the original narratives abound with stories of cartographic encounters, later popularized versions effaced them. And while the written narratives contained the story of collaboration, the images that began to dominate told a different story. Three processes were involved, including subtle re-presentations, naturalizations and erasures.

Re-presentations

It started early. Henry Schoolcraft first explored the American West in the first two decades of the nineteenth century by travelling down the White River into what is now Arkansas. He later served as geologist on an expedition led by Lewis Cass to discover the source of the Mississippi River. Schoolcraft's account of the journey was published in 1821. In 1832 he again explored the headwaters of the great river. The account was published in 1834. During this second expedition Schoolcraft drew upon the indigenous people.

> At the mouth of the Brule, a small party of the Chippewas was encountered . . . It turned out to be the family of Oza Windib, one

> of the principal Chippewas from Cass Lake. He was persuaded to return, and proved himself to be a trusty and experienced guide through the most remote and difficult parts of the journey.[1]

In the subsequent journey from Cass Lake to Itasca Lake, he relied on guides:

> Having determined to organise a select party at this lake, to explore the source of the river, measures were immediately taken to effect it. A council of the Indians was assembled, and the object declared to them. They were requested to delineate maps of the country, and to furnish the requisite number of hunting canoes and guides. Oza Windib said, 'My father, the country you are going to see is my hunting ground. I have travelled with you many days. I shall go with you farther. I will myself furnish the maps you have requested, and will guide you onward.'[2]

Before nightfall the maps were completed. The Chippewa guides took Schoolcraft's party to Itasca Lake, the source of the Mississippi. Schoolcraft and his men wanted a memorial and so felled a few trees, erected a flagstaff and placed an American flag. On the small map that accompanied the narrative, the island was referred to as Schoolcraft's Island. This was a fairly typical encounter involving local guides and

44 'Itasca Lake', *c.* 1857, from Henry Schoolcraft, *History of The Indian Tribes of The United States* . . . (Philadelphia, 1857).

the process of naming as possession. A visual record of the event was also made. Schoolcraft's original sketch was turned into a lithograph for an 1851 volume (illus. 43). In this rendition a man sits beside a tent waiting for the arrival of a party including Indian guides. This same scene was also turned into a steel engraving for an 1857 edition (illus. 44). A subtle change took place. In the later image, as Martha Sandweiss has noted, a white man is shown standing tall at the front of the party. Collaboration is turned into a subtle representation of a dominant (white) hero leading the way forward.[3]

By such subtle changes collaborations and encounters were turned into romanticized heroic missions into the wilderness.

Naturalizations

As images more than words became the most popular form of consuming the exploration narratives, the context of Native Americans representation was important. They were shown in illustrations along with depictions of scenery, flora and fauna. Set beside snakes and mammals, geological landforms and mountain scenery, flowers and trees they became integrated into the 'natural' order of things, items of the physical landscape to be studied and illustrated and less people of the social world to be treated with mutual respect. Denied full social

45 'Navajos', from H. B. Mollhausen, *Pacific Railroad Reports*, volume III (Washington, DC, 1856).

46 J. H. Richard, illustration of a snake from W. H. Emory, *Report of the United States and Mexican Boundary Survey . . .* (Washington, DC, 1857).

status, they were reduced to background scenery. Downgraded visually to the label of specimens or generic types, their humanity – and hence their active creative role in the process of exploration – was easily overlooked, ignored or forgotten.

47 F. W. Egloffstein, 'Franklin Valley', *c.* 1855, from *Pacific Railroad Reports*, volume II (Washington, DC, 1855).

Erasures

And finally there were simple erasures from the landscape. Egloffstein's 'Franklin Valley' from the middle of the nineteenth century, shows the beginning of the romantic tradition that was later to dominate depiction of the American West. Notice how the figure in the left, a useful device to indicate the sheer size of the landscape, is identifiably Native American. With bow and arrow in hand and dog beside, he contemplates the vastness. At the margins of the picture, to be sure, but still in the picture. Closer to the end of the century, an illustration of panoramic views, done around 1872 to illustrate the Hayden Survey, also incorporates figures to denote scale, but this time the Native American is replaced by two men wearing identifiable 'white' clothing (illus. 48). The Native American is slowly disappearing from the picture just as they have from the common understanding and representation of western exploration and mapping.

We need to look again at the history. The final illustration is from the Pacific Railroad surveys, done in the 1850s (illus. 49): it is by the painter John Mix Stanley from a drawing by Richard Kern. Stanley was sympathetic to the Native Americans. Kern was one of three brothers, all artists in western exploration. He was a member of an early Fremont expedition and in 1853 joined the Pacific Railroad Survey. He accompanied the party led by Gunnison and died at the hands of Paiute Indians in Utah. It is a rich irony that one of the rare

48 William Henry Jackson, 'Topographical Work', 1874, from F. V. Hayden, *Geological and Geographical Survey of the Territories* (Washington, DC, 1876).

49 J. M. Stanley, 'View of Sangre de Christo Pass', 1855–60, from *Pacific Railroad Reports*, volume VI.

sympathetic accounts of a cartographic encounter was done by an artist later killed by Native Americans. But that is what illustration 49 represents. An indigenous man raises his hand to show the way to the man on horseback. A well-worn track leads the way through the pass and across the mountains. The man on the horse looks down. From his exalted position on the horse he bends his neck forward to receive directions, guidance and information.

Appendix: Composite Journeys

Here I provide a flavour of cartographic encounters through a sampling of expedition accounts arranged chronologically. I include extracts already cited as well as several selections from expeditions that I have not discussed in this book, including the narratives of John Russell Bartlett, William Dunbar, Henry Schoolcraft and Captain Lorenzo Sitgreaves. Together they reveal and highlight a heavy reliance on indigenous trails, guides, mapmakers and geographic informants.

> I reached the land of Cariay, where I stayed to repair the ships and rest the crew . . . Two Indians brought me to Carambaru . . . They gave me the names of many places on the sea-coast where they said there was gold and goldfields too . . . In all these places, I had visited, I found the information given me true.[1]
> *Christopher Columbus, 1502*

> The two savages assured us that this was the way to the mouth of the great river of Hochelaga [St Lawrence] and the route towards Canada . . . and also that farther up the water became fresh.[2]
> *Jacques Cartier, 1534*

The circle of meale signified their Country, the circles of corne the bounds of the Sea, and the stickes his Country.[3]
John Smith, 1607

After good deliberation he began to describe mee the Countreys beyonde the Falles.[4]
John Smith, 1607

This morning two Canoes came up the River where we first found loving people, and in one of them was an old man . . . He brought another old man with him, which brought more ropes of Beades, and gave them to our Master, and shewed him all the Countrey.[5]
Robert Juet, 1609

I enquired of this king how farre this River ranne up into the Countrey, and whither it werr navigable or no, he told me it ranne a great way up, and that I might gow with my shippe, till I came to a certaine place, where the rockes ranne cleane across the River . . . I then desired him to lend me a pilott to goe up to that place, which he most willingly granted.[6]
Thomas Yong, 1634

This young man traced for us with coal, a pretty exact map, assuring us that he had been everywhere in his periagua [canoe]; that there was not down to the sea, which the Indians call the great lake, either falls or rapids. But that as this river became very broad, there were in some places sand banks and mud which barred a part of it. He also told us the name of the nations that lived on its bank, and of the rivers which it receives. I wrote them down.[7]
Louis Hennepin, 1679

But as this enterprise required a great outlay, we wished to learn whether their river was navigable to the sea, and whether other Europeans dwelt near its mouth. The Illinois replied that they accepted all our proposals, and that they would assist us as far

as they could, then they gave a description of the river Colbert or Meschasipi [Mississippi]; they told us wonders of its width, and beauty, and they assured us that the navigation was free and easy, and that there were no Europeans near its mouth.[8]
Louis Hennepin, 1679

The 9th and 10th were spent in visits, and we were informed by one of the Indians that we were not far from a great river, which he described with stick on the sand, and showed it had two branches.[9]
Henri Joutel, 1686

M. Le Fevre, an intelligent man, a native of the Illinois . . . possesses considerable knowledge of the interior of the country; he confirms the accounts we have already obtained . . . I asked M. Le Fevre for information.[10]
William Dunbar, 1805

I now prevailed upon the Chief to instruct me with rispect to the geography of the country. This he undertook very cheerfully, by delienating the river on the ground . . . he drew the river on which we now are (Lemhi River) to which he placed two branches just above us . . . he next made it discharge itself into a large river which flowed from the S.W. abut ten miles below us (Salmon River), then continued this joint stream in the same direction of this valley . . . for one days march and then encircled it to the West for 2 more days march. Here he placed a number of heaps of sand on each side which he informed me represented the vast mountains of rock.[11]
Meriwether Lewis, 1805

At a short distance from our place of encampment we passed an Indian camp . . . Well worn traces or paths lead in various directions from this spot.[12]
Edwin James, 1820

We pursued the Indian path a considerable distance.[13]
Edwin James, 1820

Having determined to organise a select party at this lake, to explore the source of the river, measures were immediately taken to effect it. A council of the Indians was assembled, and the object declared to them. They were requested to delineate maps of the country, and to furnish the requisite number of hunting canoes and guides. Oza Windib said, 'My father, the country you are going to see is my hunting ground. I have travelled with you many days. I shall go with you farther. I will myself furnish the maps you have requested, and will guide you onward.'[14]
Henry Schoolcraft, 1832

After supper we sat down on the grass, and I placed a sheet of paper between us, on which they traced rudely, but with a certain degree of relative truth, the watercourses of the country, which lay between their villages, and us, and of which I desired to have some information.[15]
John Charles Fremont, 1842

They [Native Americans] made on the ground a drawing of the river which they represented as issuing from another lake in the mountain three or four days distant, in a direction a little west of south; beyond which they drew a mountain; and further still two rivers; on one of which they told us that people like ourselves traveled.[16]
John Charles Fremont, 1844

We now placed before them a map of the country, which had been made out at Bent's Fort by Tah-Kai-Buhl . . . Quite a council was called to decide whether or not Buffalo Creek runs into Red River or into the 'Goo-al-pal' as represented on the map. A clean sheet of paper was produced, and the map drawn according to their directions. Tah-Kai-Buhl's map was corrected and the relative portions of the various topographical features were preserved in a surprisingly exact manner, when we consider that it covers an area of about 800 miles in length.[17]
James William Abert, 1845

Left at the lodges a small present of tobacco, handkerchiefs, and knives for the purpose of conciliating the Indians and inducing them to hold some intercourse with us, by which means we hoped to obtain useful information in regard to the route.[18]
Captain L. Sitgreaves, 1851

Taking an old Comanche trail this morning, I followed it to a narrow defile in the mountains, which led me up through a very tortuous and rocky gorge, where the well-worn path indicated that it had been travelled for many years.[19]
Captain Randolph Marcy, 1852

The Indians told me we had better remain where we were, as there was no grass near the river . . . After crossing a deep arroyo of sand, which is filled by the river in its flood, and pushing our way through thick underbrush of willows, we at length reached the bank of the river, when I found the statement of the Indians too true . . . there was not a blade of grass to be see, and, what was worse, *the Gila was dry!*[20]
John Russell Bartlett, 1853

We crossed during the afternoon the trail leading toward the Little Blackfoot fork, which is the main Flathead trail across the mountains.[21]
Lt John Mullan, 1854

In making inquiries of the Indians we were cautious not to incline their minds to any preconceived notions of our own by asking leading questions. But we traced in the sand a depression to denote the valley of the Colorado, and in the middle represented the meandering river which they recognized. Heaps of sand piled on each side, and called sierras, they understood to denote mountains. Our trail was then marked out, indicating the camps and the springs, and the mountains crossed from the Mojave villages to the flowing water of Rio Mojave. Thence we represented the valley of that river by a channel scooped from the sand, indicating where there was water, and where there was

none. Then we inquired of the Indians whether or not the channel was lost in the lake. The guide understood the question, and instantly with his hand cut a passage through the heaps of sand that had been piled upon the ground-map for mountains around the lake, representing a continuous valley to the arroyo by which we haft the Colorado. Pointing along the line traced, he said it was a smooth arroyo, containing no water upon the surface, nor obstructed by hills, but the valley was wide, level and sandy throughout. When asked if the distance was great, he replied that it was; and counting with his fingers, he indicated ten as the number of days required to travel it with our wagons.[22]
Captain A. Whipple, 1854

REFERENCES

1 | Creation Myths and Cartographic Encounters

1 *Montreal Official Tourist Guide* (Montreal, 2006–7), p. 19.

2 I have drawn heavily on my chapter on the Western film tradition in John Rennie Short, *Imagined Country* (Syracuse, 1991, repr. 2006), pp. 178–96.

3 The quotation is from J. H. Merrell, 'Indian History during the English Colonial Era', in *A Companion to Colonial America*, ed. D. Vickers (Oxford, 2003), p. 118. The two standard works are F. Jennings, *The Invasion of America: Indians, Colonialism and the Cant of Conquest* (Chapel Hill, NC, 1975) and R. White, *The Middle Ground: Indians, Empires and Republics in the Great Lakes Region, 1650–1815* (New York, 1991). For more recent works that explore this middle ground see R. Bourne, *Gods of War, Gods of Peace: How the Meeting of Native and Colonial Religions Shaped Early America* (New York, 2002); K. O. Kupperman, *Indians and English: Facing Off in Early America* (Ithaca, NY, and London, 2000); M. L. Oberg, *Dominion and Civility: English Imperialism and Native America, 1585–1685* (Ithaca, NY, and London, 1999); J. A. Sokolow, *The Great Encounter: Native Peoples and European Settlers in the Americas, 1492–1800* (Armonk, NY, 2003).

4 B. J. Harley, *The New Nature of Maps* (Baltimore, MD, and London, 2001).

5 G. M. Lewis, ed., *Cartographic Encounters: Perspectives on Native American Mapmaking and Map Use* (Chicago, IL, 1998).
6 W. H. Goetzmann and G. Williams, eds, *The Atlas of North American Exploration* (New York, 1992), p. 26.
7 I have used the following edition: J. Cartier, *The Voyages of Jacques Cartier* ed. Henry Percival Biggar, intro. Ramsay Cook (Toronto, 1993).
8 Ibid., p. 44. The raising of crosses was a typical act of claiming of territory and an effective dispossession of indigenous land rights. A lack of European religion signified a land ripe for appropriation. The ritual of cross raising was both a religious symbol and a political act: see P. Seed, *Ceremonies of Possession in Europe's Conquest of the New World, 1492–1640* (Cambridge, 1995); G. Sabo III, 'Rituals of Encounter: Interpreting Native American Views of European Explorers', in Jeannie M. Whayne, ed., *Cultural Encounters in Early America: Native Americans and Europeans in Arkansas* (Fayetteville, AR, 1995), pp. 76–87.
9 Cartier, *Voyages of Jacques Cartier*, p. 26.
10 Ibid., p. 44–5.
11 Ibid., p. 75.
12 Quoted in J. M. Cohen, ed. and trans., *The Four Voyages of Christopher Columbus* (Harmondsworth, 1969), p. 287.
13 Quoted in W. Irving, *The Life and Voyages of Christopher Columbus*, vol. II (Whitefish, MT, [1892] 2004), p. 186. The scene as described by Irving draws heavily upon contemporary accounts by members of the expedition, including Diego Mendez and Columbus's son Hernando Colon. For a recent reworking of these sources into a narrative account see M. Dugard, *The Last Voyage of Columbus* (New York and Boston, MA, 2005).
14 Cartier, *Voyages of Jacques Cartier*, p. 52.
15 H. Warre, *Sketches in North America and the Oregon Territory* (London, 1848).
16 C. G. Calloway, *New Worlds for All: Indians, Europeans and the Remaking of Early America* (Baltimore, MD, 1997), p. 198.

2 | Amerindian Mappings

1 W. P. Cumming, *The Southeast in Early Maps*, 3rd edn, revised and

enlarged by Louis de Vorsey, Jr (Chapel Hill, NC, and London, 1998). See also G. M. Lewis, ed., *Cartographic Encounters: Perspectives on Native American Mapmaking and Map Use* (Chicago, IL, 1998); M. Warhus, *Another America: Native American Maps and the History of Our Land* (New York, 1997) and chapters 4, 5, 6 and 7 in D. Woodward and G. M. Lewis, eds, *The History of Cartography*, vol. 2, Bk 3: *Cartography in the Traditional African, American, Arctic, Australian, and Pacific Societies* (Chicago, 1998).

2 Barabara E. Mundy, 'Mesoamerican Cartography', in *Cartography*, ed. Woodward and Lewis, pp. 183–256.

3 Copies of the various maps mentioned in this chapter are available in J. R. Short, *The World through Maps* (Toronto, 2003).

4 W. G. Gartner, 'Mapmaking in the Central Andes', in *Cartography*, ed. Woodward and Lewis, pp. 257–300.

5 Cumming, *The Southeast in Early Maps*, pp. 108–10.

6 C. Nicholl, *The Creature in the Map* (London, 1995).

7 S. C. Vehik, 'Onate's Expedition to the Southern Plains: Routes, Destinations, and Implications for Late Prehistoric Cultural Adaptions', *Plains Anthropologist*, XXXI (1986), pp. 13–33.

8 E. Arber, ed., *Travels and Works of Captain John Smith* (New York, 1910), pp. 76–7.

9 Ibid., p. 399.

10 F. W. Gleach, *Powhatan's World and Colonial Virginia: A Conflict of Cultures* (Lincoln, NE, 1997).

11 I have drawn upon N. Philbrick, *Mayflower: A Story of Courage, Community and War* (New York, 2006), p. 108.

3 | Encounters in a Settled Land

1 C. Martin, *Keepers of the Game* (Berkeley and Los Angeles, CA, 1978). For a more recent contribution see P. Nadasdy, 'Transcending the Debate over the Ecologically Noble Indian: Indigenous Peoples and Environmentalism', *Ethnohistory*, LII (2005), pp. 291–331.

2 Quoted in P. Nabokov, ed., *Native American Testimony*, revd edn (New York 1999), p. 21.

3 Ibid., p. 25.

4 Charles Mann's *1491* (New York, 2005) contains a full discussion of

the academic works that estimate pre-contact population numbers.

5 Quoted in J. F. Jameson, ed., *Narratives of New Netherland, 1609–1664* (New York, 1990), p. 23.

6 Ibid., p. 24.

7 The map was found in a case in the Royal Archives in The Hague by J. R. Brodhead, who signed himself 'Agent of the State of New York'. Brodhead had been charged with obtaining copies of documents in foreign archives relating to the history of the state of New York. There was no mark on it to ascertain its exact date, but it was found along with a document dated 1616, which has led to the map being dated in some libraries, such as the State Library of New York, as 1616. The map found in the case is probably a copy of Block's original map.

8 The map is available at the Illinois State Museum website, www.museum.state.il.us/muslink/nat_amer/post/htmls/popups/ic_map.html (accessed 30 July 2008).

9 In 1682, according to the remembrance of Father Zenobius Membre, 'on the ninth of April, with all possible solemnity, we performed the ceremony of planting the cross and raising the arms of France. After we chanted the hymn of the church, "Vexilla Regis" and the Te Deum, the Sieur de la Salle in the name of his majesty took possession of that river, of all the rivers that enter it and of all the country watered by them. An authentic act was drawn up, signed by all of us there, and, amid volley from all our muskets, a leaden plate subscribed with the arms of France and the names of those who had just made the discovery was deposited in the earth.' Quoted from I. J. Cox, ed., *The Journeys of René Robert Cavelier, Sieur de La Salle* (New York, 1905), p. 145.

10 *Description of Louisiana* was first published in France in 1683. I have used the translation edited by John Gilmary Shea (New York, 1880) pp. 183–4.

11 Ibid., p. 256.

12 Ibid., p. 140.

13 Ibid., p. 161.

14 Ibid., pp. 162–3.

15 Cox, *Journeys of René Robert Cavelier*, vol. II, p. 189.

16 J. R. Short, *Representing the Republic* (London, 2001).

17 J. B. Harley, 'New England Cartography and the Native Americans',

in E. W. Baker and E. A. Churchill, eds, *American Beginnings: Exploration, Culture and Cartography in the Land of Norumbega* (Lincoln, NE, and London, 1994); L. De Vorsey, 'Amerindian Contributions to the Mapping of North America: A Preliminary View', *Imago Mundi*, XXX (1978), pp. 71–8; G. M. Lewis, ed., *Cartographic Encounters: Perspectives on Native American Mapmaking and Map Use* (Chicago, IL, 1998).

4 | Landing in a Strange Land

1 E. Arber, ed., *Travels and Works of Captain John Smith* (New York, 1910), p. 6.
2 Ibid., p. 7.
3 Ibid., p. 14.
4 Ibid., pp. 16–17.
5 Ibid., p. 19.
6 The telling juxtaposition is presented in E. C. Papenfuse and J. M. Coale III, *The Maryland State Archives Atlas of Historical Maps of Maryland 1608–1908* (Baltimore MD, and London, 2003), pp. 22–3. The map is available on the Bodleian Library, Oxford, website at www.rsl.ox.ac.uk/guides/maps/virginia.jpg (accessed 30 July 2008).
7 Arber, *Travels and Works of Captain John Smith*, p. 55.
8 'Relation of Captain Thomas Yong, 1634', in A. C. Myers, ed., *Narratives of Early Pennsylvannia, West New Jersey and Delaware* (New York, 1911), p. 37.
9 Ibid., p. 38.
10 Ibid., p. 41.

5 | Surveying the West

1 Amongst the many volumes see C. Gilman, *Lewis and Clark: Across the Divide* (Washington and London, 2003). For a sample of more detailed studies see J. P. Ronda, *Lewis and Clark among the Indians* (Lincoln, NE, 1984) and '"A Chart in His Way": Indian cartography and the Lewis and Clark Expedition', in F. C. Luebke, F. W. Kaye and G. E. Moulton, eds, *Mapping the North American Plains* (Norman, OK, and London, 1987), pp. 81–92; T. P. Slaughter, *Exploring Lewis and Clark* (New York, 2003). There are also two

very good online exhibitions, at the Missouri Historical Society website, www.lewisandclarkexhibit.org (accessed 31 July 2008) and at the Library of Congress website, www.loc.gov/exhibits/lewisandclark/lewisandclark.html (accessed 31 July 2008).

2 B. DeVoto, ed., *The Journals of Lewis and Clark* (Boston, MA, 1953), p. 101.

3 Ibid., p. 159.

4 Ibid., pp. 365–6.

5 Ibid., p. 400.

6 Ibid., p. 77.

7 Ibid., pp. 210–11.

8 F. Bergson, ed., *The Journals of Lewis and Clark* (New York, 1989), p. 284.

9 Many of the Lewis and Clark maps drawn from indigenous informants are in the Beineke Library at Yale University.

10 The story is available at the National Public Radio website, www.npr.org/templates/story/story.php?storyId=1748763 (accessed 31 July 2008), while the map is available at www.npr.org/programs/atc/features/2004/mar/lewisclark/map.html.

11 See L. Y. Jones, *William Clark and the Shaping of the West* (New York, 2004).

6 | Expedition into the 'Desert'

1 Among recent work on the role of Humboldt in establishing the model for scientific exploration in the nineteenth century and the way we see the world today, see A. Sachs, *The Humboldt Current: Nineteenth-Century Exploration and the Roots of American Environmentalism* (New York, 2006), also G. Helferich, *Humboldt's Cosmos: Alexander von Humboldt and the Latin American Journey that Changed the Way We See the World* (New York, 2004).

2 E. James, *An Account of an Expedition from Pittsburgh to the Rocky Mountains* (Philadelphia, PA, [1823] 1972), p. 4.

3 The letters are held as *Edwin James Letterbook*, Western Americana Collection, Beinecke Rare Book and Manuscript Library, Yale University. The quotes are taken from letters in this collection.

4 James, *An Account of an Expedition*, p. 158.

5 K. Haltman, 'Private Impressions and Public Views: Titian Ramsey

Peale's Sketchbooks from the Long Expedition', *Yale University Gallery Bulletin* (Spring 1989), pp. 39–53. See also Haltman, *Looking Close and Seeing Far: Samuel Seymour, Titian Ramsey Peale and the Art of the Long Expedition* (University Park, PA, 2007).

6 E. Cohen, *Mapping the West: America's Westward Movement 1524–1890* (New York, 2002), p. 108. Information on the Long map is available at the New York Public Library website, www.nypl.org/west/hw_explor1.shtml (accessed 31 July 2008), and the McFarlin Library, University of Tulsa website, www.lib.utulsa.edu/speccoll/collections/maps/long/index.htm (accessed 31 July 2008).

7 James, *An Account of an Expedition*, p. 392

8 Ibid., p. 476.

9 Ibid., p. 477.

10 Ibid., p. 478.

11 Ibid., p. 483.

12 Ibid., p. 484.

13 Ibid., p. 486.

7 | Fremont and Tah-Kai-Buhl

1 The two biographies are T. Chaffin, *Pathfinder: John Charles Fremont and the Course of American Empire* (New York, 2002) and D. Roberts, *A Newer World: Kit Carson, John C. Fremont and the Claiming of the American West* (New York, 2000). The entry is in H. R. Lamar, ed., *The New Encyclopedia of the American West* (New Haven, CT, and London, 1998), pp. 401–2.

2 The standard work is W. H. Goetzmann, *Army Exploration in the American West* (New Haven, CT, 1959) and his later *Exploration and Empire: The Explorer and the Scientist in the Winning of the American West* (New York, 1966). Also very useful and concise is J. P. Ronda, *Beyond Lewis and Clark: The Army Explores the West* (Tacoma, WA, 2003). See also D. Reinhartz and G. D. Saxon, eds, *Mapping and Empire: Soldier-Engineers on the Southwestern Frontier* (Austin, TX, 2005).

3 I discuss the meaning of wilderness in a previous work: J. R. Short, *Imagined Country* (Syracuse, NY, 1991), pp. 3–27.

4 Le Conte Papers, American Philosophical Society, Philadelphia, reference B L 493.3. The letters from Hammond are dated June 1854,

29 June 1856, 24 August 1856, 18 January 1857 and 23 February 1857. From the tone of the letters it is clear that Hammond is lonely on the frontier and missing the intellectual life of the city and its scientists.

5 J. N. Nicollet, *Report Intended to Illustrate Map of the Hydrographical Basin of the Upper Mississippi* (Washington, 1845), p. 3.
6 See E. C. Bray and M. C. Bray, *Joseph N. Nicollet and the Plains and Prairies* (St Paul, MN, 1993).
7 Nicollet, *Report Intended to Illustrate Map of the Hydrographical Basin of the Upper Mississippi*, p. 103.
8 Ibid., p. 45.
9 Quoted in D. Jackson and M. L. Lee, eds, *The Expeditions of John Charles Fremont* (Urbana, IL, and London, 1970), vol. I, p. 185.
10 Ibid., p. 233.
11 Ibid., p. 610.
12 Ibid., vol. II, p. 7.
13 It has been republished as J. W. Abert, *Through the Country of the Comanche Indians in the Fall of 1845*, ed. J. Galvin (San Francisco, 1970).
14 Ibid., p. 2.
15 Ibid., p. 34.
16 Ibid., p. 51.
17 *Blackwood Edinburgh Magazine* (1848), p. 440.
18 See D. C. Comer, *Ritual Ground: Bent's Old Fort, World Formation and the Annexation of the Southwest* (Berkeley and Los Angeles, 1996).

8 | 'Warren's Map'

1 Quoted in W. H. Goetzmann and G. Williams, eds, *The Atlas of North American Exploration* (New York, 1992), p. 169.
2 An illustrated account is available in H. J. Viola and C. M. Margolis, eds, *Magnificent Voyagers: United States Exploring Expedition 1838–1842* (Washington, DC, 1985). A standard work is W. Stanton, *The Great United States Exploring Expedition of 1838–1842* (Berkeley, CA, 1975). A more recent popularization is N. Philbrick, *Sea of Glory: America's Voyage of Discovery, the US Exploring Expedition, 1838–1842* (New York, 2003).
3 C. Wilkes, *United States Exploring Expedition, 1838–1842*

(Philadelphia, 1849), p. 568.

4 Henry Eld Papers MSS 161, Western Americana Collection, Beinecke Library, Yale University.

5 J. H. Simpson, *Journal of a Reconnaissance from Sante Fe, New Mexico to the Navajo Country* (Philadelphia, PA, 1852).

6 Ibid., pp. 80–81.

7 R. Marcy, *Exploration of the Red River of Louisiana in the Year 1852* (Washington, dc, 1853), p. 18.

8 Ibid., p. 29.

9 Ibid., p. 21.

10 Ibid., p. 75.

11 It was also hugely expensive. While the surveys themselves cost close to half a million dollars the publication costs of PRR alone amounted to $1 million; in today's money that would be $23 million using the Consumer Price Index or a staggering $2.8 billion using the relative share of GDP. Between 1840 and 1860 the US government spent almost a quarter of the federal budget funding western explorations and the subsequent publications. Some politicians accepted the huge cost. Senator James Harlan of Iowa praised the illustrations, 'Every unusual swell of land, every . . . unanticipated gorge in the mountains has been displayed in a beautiful picture. Every bird that flies in the air over the immense region, and every beast that traverses the plains and mountains, every fish that swims in its lakes and rivers, every reptile that crawls, every insect that buzzes in the summer breeze has been displayed in the highest style of art, and in most brilliant colors.' Cited in R. E. Taft, *Artists and Illustrators of the Old West, 1850–1900* (New York, 1953), p. 360.

Representative Davis of Maryland, in the contrast you would expect from someone on the eastern seaboard, argued in the *Congressional Globe* of 4 April 1856 that the reports were a money pit. 'Elegant views of scenery, disquisitions and personal incidents, descriptions of the red man, which crowd the pages of all the reconaissances which have been published during the last four or five years, have no business in Government publications, and ought not to be sanctioned.'

See also M. Sandweiss, 'The Public Life of Western Art', in J. D. Prown, ed., *Discovered Lands, Invented Pasts* (New Haven, CT, 1992), pp. 117–33; R. Tyler, 'Illustrated Government Publications Relating to the American West, 1843–1863' in E. C. Carter, ed., *Surveying the Record: North American Scientific Explorations to 1930* (Philadelphia, PA, 1999), pp. 147–72.

12 US Army *Reports of Explorations and Surveys to Ascertain the Most Practicable and Economic Route for a Railroad from the Mississippi to the Pacific Ocean* (Washington, DC, 1855–60), vol. I, p. 309.

13 Ibid., p. 311.

14 Ibid., p. 345.

15 Ibid., vol. II, p. 6.

16 Ibid., vol. II, p. 1.

17 Ibid., vol. III; the quotes are, in order, taken from pp. 14, 16, 17, 17, 18, 22 and 126.

18 G. K. Warren, *Explorations in the Dacota Country in the Years 1855–'56–'57* (Washington, 1875), p. 17.

19 Ibid., p. 7.

20 Ibid., p. 57.

9 | Closing the Frontier in the West

1 G. Wheeler, *Report upon United States Geographical Surveys West of the 100th Meridian* (Washington, DC, 1889), vol. I, p. 34.

2 The letters are held in the Western American Collection, Beinecke Library, Yale University: Geographical Surveys West of the 100th Meridian Papers WA MSS S-744/ Box 1.

3 Ibid., folder 45.

4 Wheeler, *Report upon United States Geographical Surveys*, p. 35.

10 | Cartographic Encounters in Australia

1 C. Sturt, *Narrative of an Expedition into Central Australia* (London, 1849), p. 34.

2 Ibid., p. 104.

3 Ibid., pp. 118–19.

4 Ibid., p. 124.

5 Ibid., p. 125.

6 Ibid., p. 311.

7 T. L. Mitchell, *Three Expeditions into the Interior of Eastern*

Australia, Volume Two (London, 1838), p. 3.
8 Ibid., p. 30.
9 Ibid., p. 162.
10 E. J. Eyre, *Journals of Expedition and Discovery* (London, 1845), p. 212.
11 Ibid., p. 217.
12 Ibid., pp. 279–80.
13 L. Leichhardt, *Journal of an Overland Expedition in Australia* (London, 1847), p. 503.
14 J. M. Stuart, *Explorations Cross the Continent of Australia 1861–62* (Adelaide, [1863] 1963), p. 21.
15 Ibid., p. 22.
16 Ibid., p. 38.
17 E. Giles, *Geographic Travels in Central Australia from 1872 to 1874* (Melbourne, 1875), p. 43.
18 Ibid., p. 44.
19 Ibid., p. 136.
20 T. L. Mitchell, 'Account of the Recent Exploring Expedition to the Interior of Australia', *Journal of the Royal Geographical Society*, VII (1837), pp. 271–85.

11 | Journey's End

1 H. Schoolcraft, *Narrative of an Expedition through the Upper Mississippi to Itasca Lake* (New York, 1834), p. 21.
2 Ibid., p. 40.
3 M. Sandweiss, 'The Public Life of Western Art', in J. D. Prown, ed., *Discovered Lands, Invented Pasts* (New Haven, CT, 1992).

Appendix

1 Quoted in J. M. Cohen, ed. and trans., *The Four Voyages of Christopher Columbus* (Harmondsworth, 1969), p. 287.
2 J. Cartier, *The Voyages of Jacques Cartier* (Toronto, 1993), p. 44.
3 E. Arber, *Travels and Works of Captain John Smith* (New York, 1910), p. 399.
4 Ibid., p. 19.
5 J. F. Jameson, ed., *Narratives of New Netherland, 1609–1664* (New

York, 1990), p. 24.

6 A. C. Myers, ed., *Narratives of Early Pennsylvania, West New Jersey and Delaware* (New York, 1911), p. 37.

7 J. G. Shea, ed., *Description of Louisiana* (New York, 1880), pp. 183–4.

8 Ibid., pp. 162–3.

9 I. C. Cox, ed., *The Journeys of René Robert Cavelier, Sieur de La Salle* (New York, 1905), p. 189.

10 W. Dunbar, *Documents Relating to the Purchase and Exploration of Louisiana* (Boston, MA, and New York, 1904), pp. 158–68.

11 B. DeVoto, ed., *The Journals of Lewis and Clark* (Boston, MA, 1953), pp. 210–11.

12 E. James, *An Account of an Expedition from Pittsburgh to the Rocky Mountains* (Philadelphia, PA, 1823), p. 476.

13 Ibid., p. 484.

14 H. Schoolcraft, *Narrative of an Expedition through the Upper Mississippi to Itasca Lake* (New York, 1834), p. 40.

15 D. Jackson and M. L. Lee, eds, *The Expeditions of John Charles Fremont* (Urbana, IL, and London, 1970), vol. 1, p. 185.

16 Ibid., p. 610.

17 J. W. Abert, *Through the Country of the Comanche Indians in the Fall of 1845*, ed. J. Galvin (San Francisco, CA, 1970), p. 51.

18 L. Sitgreaves, *Report of an Expedition down the Zuni and Colorado Rivers* (Washington, DC, 1854), p. 11.

19 R. Marcy, *Exploration of the Red River of Louisiana in the Year 1852* (Washington, DC, 1853), p. 21.

20 J. R. Bartlett, *Personal Narrative of Explorations and Incidents* (New York, 1854), vol. II, pp. 214–15.

21 US Army, *Reports of Explorations and Surveys to Ascertain the Most Practicable and Economic Route for a Railroad from the Mississippi to the Pacific Ocean* (Washington, DC, 1855–60), vol. III, p. 345.

22 Ibid., p. 126.

BIBLIOGRAPHY

Abert, J. W., *Through the Country of the Comanche Indians in the Fall of 1845*, ed. J. Galvin (San Francisco, CA, 1970)

Arber, E., ed., *Travels and Works of Captain John Smith* (New York, 1910)

Bartlett, J. R., *Personal Narrative of Explorations and Incidents* (New York, 1854)

Bergon, F., ed., *The Journals of Lewis and Clark* (New York, 1989)

Bourne, R., *Gods of War, Gods of Peace: How the Meeting of Native and Colonial Religions Shaped Early America* (New York, 2002)

Bray, E. C., and M. C. Bray, eds, *Joseph N. Nicollet on the Plains and Prairies* (St Paul, MN, 1993)

Calloway, C. G., *New Worlds for All: Indians, Europeans and the Remaking of Early America* (Baltimore, MD, 1997)

Cartier, J., *The Voyages of Jacques Cartier* (Toronto, 1993)

Chaffin, T., *Pathfinder: John Charles Fremont and the Course of American Empire* (New York, 2002)

Cohen, E., *Mapping the West: America's Westward Movement 1524–1890* (New York, 2002)

Cohen, J. M., ed. and trans., *The Four Voyages of Christopher Columbus* (Harmondsworth, 1969)

Comer, D. C., *Ritual Ground: Bent's Old Fort, World Formation, and the Annexation of the Southwest* (Berkeley and Los Angeles, 1996)

Cox, I. J., ed., *The Journeys of René Robert Cavelier, Sieur de La Salle* (New York, 1905)
Cumming, W. P., *The Southeast in Early Maps*, 3rd edn, rev. Louis de Vorsey Jr (Chapel Hill, NC, and London, 1998)
De Vorsey, L., 'Amerindian Contributions to the Mapping of North America: A Preliminary View', *Imago Mundi*, XXX (1978), pp. 71–8
DeVoto, B., ed., *The Journals of Lewis and Clark* (Boston, MA, 1953)
Dugard, M., *The Last Voyage of Columbus* (New York and Boston, MA, 2005)
Dunbar, W., *Documents Relating to the Purchase and Exploration of Louisiana* (Boston, MA, and New York, 1904)
Eyre, E. J., *Journals of Expeditions of Discovery* (London, 1845)
Gartner, W. G., 'Mapmaking in the Central Andes', in D. Woodward and G. M. Lewis, eds, *Cartography in the Traditional African, American, Arctic, Australian, and Pacific Societies* (Chicago, IL, 1998)
Giles, E., *Geographic Travels in Central Australia from 1872 to 1874* (Melbourne, 1875)
Gilman, C., *Lewis and Clark: Across the Divide* (Washington, DC, and London, 2003)
Gleach, F. W., *Powhatan's World and Colonial Virginia: A Conflict of Cultures* (Lincoln, NE, 1997)
Goetzmann, W. H., *Army Exploration in the American West* (New Haven, CT, 1959)
—, *Exploration and Empire: The Explorer and the Scientist in the Winning of the American West* (New York, 1966)
—, and G. Williams, eds, *The Atlas of North American Exploration* (New York, 1992)
Haltman, K., 'Private Impressions and Public Views: Titian Ramsay Peale's Sketchbooks from the Long Expedition', *Yale University Gallery Bulletin* (Spring 1989), pp. 39–53
—, *Looking Close and Seeing Far: Samuel Seymour, Titian Ramsay Peale and the Art of the Long Expedition, 1818–1823* (University Park, PA, 2007)
Harley, J. B., 'New England Cartography and the Native Americans', in E. W. Baker and E. A. Churchill, eds, *American Beginnings: Exploration, Culture and Cartography in the Land of Norumbega* (Lincoln, NE, and London, 1994)
—, *The New Nature of Maps* (Baltimore, MD, and London, 2001)

Helferich, G., *Humboldt's Cosmos: Alexander von Humboldt and the Latin American Journey that Changed the Way We See the World* (New York, 2004)
Irving, W., *The Life and Voyages of Christopher Columbus*, vol. 2 (Whitefish, MT, [1892] 2004)
Jackson, D. and M. L. Lee, eds, *The Expeditions of John Charles Fremont* (Urbana, IL, and London, 1970)
James, E., *Account of an Expedition from Pittsburgh to the Rocky Mountains* (Philadelphia, PA, 1823)
Jameson, J. F., ed., *Narratives of New Netherland, 1609–1664* (New York, 1990)
Jennings, F., *The Invasion of America: Indians, Colonialism and the Cant of Conquest* (Chapel Hill, NC, 1975)
Jones, L. Y., *William Clark and the Shaping of the West* (New York, 2004)
Kupperman, K. O., *Indians and English: Facing Off in Early America* (Ithaca, NY, and London, 2000)
Lamar, H. R., ed., *The New Encyclopedia of the American West* (New Haven, CT, and London, 1998)
Leichhardt, L., *Journal of an Overland Expedition in Australia* (London, 1847)
Lewis, G. M., ed., *Cartographic Encounters: Perspectives on Native American Mapmaking and Map Use* (Chicago, IL, 1998)
Mann, C., *1491* (New York, 2005)
Marcy, R., *Exploration of the Red River of Louisiana in the Year 1852* (Washington, DC, 1853)
Martin, C., *Keepers of the Game: Indian–Animal Relationships and the Fur Trade* (Berkeley and Los Angeles, 1978)
Merrell, J. H., 'Indian History during the English Colonial Era', in D. Vickers, ed., *A Companion to Colonial America* (Oxford, 2003)
Mitchell, T. L., 'Account of the Recent Exploring Expedition to the Interior of Australia', *Journal of The Royal Geographical Society*, VII (1837), pp. 271–85.
—, *Three Expeditions into the Interior of Eastern Australia, Volume Two* (London, 1838)
Montreal Official Tourist Guide 2006–2007 (Montreal, 2007)
Myers, A. C., ed., *Narratives of Early Pennsylvania, West New Jersey and Delaware* (New York, 1911)
Nabokov, P., ed., *Native American Testimony*, revd edn (New York 1999)

Nadasdy, P., 'Transcending the Debate over the Ecologically Noble Indian: Indigenous Peoples and Environmentalism', *Ethnohistory*, LII (2005), pp. 291–33

Nicholl, C., *The Creature in the Map* (London, 1995)

Nicollet, J. N., *Report Intended to Illustrate a Map of the Hydrographical Basin of the Upper Mississippi* (Washington, 1845)

Oberg, M. L., *Dominion and Civility: English Imperialism and Native America, 1585–1685* (Ithaca, NY, and London, 1999)

Papenfuse, E. C. and J. M. Coale III, *The Maryland State Archives Atlas of Historical Maps of Maryland 1608–1908* (Baltimore, MD, and London, 2003)

Philbrick, N., *Sea of Glory: America's Voyage of Discovery, the U.S. Exploring Expedition, 1838–1842* (New York, 2003)

—, *Mayflower: A Story of Courage, Community and War* (New York, 2006)

Reinhartz, D. and Saxon, G. D., eds, *Mapping and Empire: Soldier-Engineers on the Southwestern Frontier* (Austin, TX, 2005)

Roberts, D., *A Newer World: Kit Carson, John C. Fremont and the Claiming of the American West* (New York, 2000)

Ronda, J. P., *Lewis and Clark among the Indians* (Lincoln, NE, 1984)

—, *Beyond Lewis and Clark: The Army Explores the West* (Tacoma, WA, 2003)

—, '"A Chart in His Way": Indian Cartography and the Lewis and Clark Expedition', in F. C. Luebke, F. W. Kaye and G. E. Moulton, eds, *Mapping The North American Plains* (Norman, OK, and London, 1987)

Sabo III, G., 'Rituals of Encounter: Interpreting Native American Views of European Explorers', in Jeannie M. Whayne, ed., *Cultural Encounters in Early America: Native Americans and Europeans in Arkansas* (Fayetteville, AR, 1995)

Sachs, A., *The Humboldt Current: Nineteenth-Century Exploration and the Roots of American Environmentalism* (New York, 2006)

Sandweiss, M., 'The Public Life of Western Art', in J. D. Prown, ed., *Discovered Lands, Invented Pasts* (New Haven, CT, 1992), pp. 117–33

Schoolcraft, H., *Narrative of an Expedition through the Upper Mississippi to Itasca Lake* (New York, 1834)

Seed, P., *Ceremonies of Possession in Europe's Conquest of the New World, 1492–1640* (Cambridge, 1995)

Shea, J. G., ed., *Description of Louisiana* (New York, 1880)
Short, J. R., *Imagined Country* (Syracuse, NY, 1991)
—, *Representing the Republic* (London, 2001)
—, *The World through Maps* (Toronto, 2003)
Simpson, J. H., *Journal of a Military Reconnaissance from Sante Fe, New Mexico to the Navajo Country* (Philadelphia, PA, 1852)
Sitgreaves, L., *Report of an Expedition down the Zuni and Colorado Rivers* (Washington, DC, 1854)
Slaughter, T. P., *Exploring Lewis and Clark* (New York, 2003)
Sokolow, J. A., *The Great Encounter: Native Peoples and European Settlers in the Americas, 1492–1800* (Armonk, NY, 2003)
Stanton, W., *The Great United States Exploring Expedition of 1838–1842* (Berkeley, CA, 1975)
Stuart, J. M., *Explorations Across the Continent of Australia 1861–62 (*Adelaide, [1863] 1963)
Sturt, C., *Narrative of an Expedition into Central Australia* (London, 1849)
Tyler, R., 'Illustrated Government Publications Relating to the American West, 1843–1863', in E. C. Carter, ed., *Surveying the Record: North American Scientific Explorations to 1930* (Philadelphia, PA, 1999), pp. 147–72
US Army *Reports of Explorations and Surveys to Ascertain the Most Practicable and Economic Route for a Railroad from the Mississippi River to the Pacific Ocean* (Washington, DC, 1855–60)
Vehik, S. C., 'Oñate's Expedition to the Southern Plains: routes, destinations, and implications for late prehistoric cultural adaptions', *Plains Anthropologist*, 31 (1986), pp. 13–33
Viola, H. J., and C. M. Margolis, eds, *Magnificent Voyagers: United States Exploring Expedition 1838–1842* (Washington, DC, 1985)
Warhus, M., *Another America: Native American Maps and the History of Our Land* (New York, 1997)
Warre, H., *Sketches in North America and the Oregon Territory* (London, 1848)
Warren, G. K., *Explorations in The Dacota Country in the Years 1855–'56–'57* (Washington, DC, 1875)
Wheeler, G., *Report upon United States Geographical Surveys West of the 100th Meridian* (Washington, DC, 1889)
White, R., *The Middle Ground: Indians, Empires and Republics in the Great Lakes Region, 1650-1815* (New York, 1991)

Wilkes, C., *United States Exploring Expedition, 1838–1842* (Philadelphia, PA, 1849)

Woodward, D., and G. M. Lewis, eds, *The History of Cartography*, vol.1, Bk. 3: *Cartography in the Traditional African, American, Arctic, Australian, and Pacific Societies* (Chicago, IL, 1998)

ACKNOWLEDGEMENTS

My greatest debts were run up during 2001 when I had the privilege to be the Alexander O. Vietor Fellow in Cartography at the Beinecke Rare Book and Manuscript Library, Yale University. George Miles, the William Robertson Coe Curator of Western Americana, was enormously helpful and accommodating. Fred Musto, then the curator of maps at Yale's Sterling Memorial Library, also provided access to the map collection. I also used the resources of the American Philosophical Library in Philadelphia during a number of visits and, yet again, Roy Goodman was a model of help and enthusiasm. Since 2002 I have made regular use of the Library of Congress, the Folger Library in Washington, DC, and the Kuhn Library at the University of Maryland Baltimore County. The Australian material presented in Chapter 11 was the result of annual visits from 1997 to 2004 to the Mitchell Library of the State Library of New South Wales in Sydney.

This book has been slowly fermenting since 1996 when I first started to write about cartography in the early Republic. It was pushed aside as I moved houses, jobs and zip codes and pursued my other interest of urban research. But it was kept alive by invitations to give talks in various venues including the Maps and Society Lecture Series in 2002 at the Warburg Institute in London, the Mattson-New York Times Annual Lecture at the Osher Map Library at University of Southern Maine in 2004, the Medieval and Renaissance Colloquia at the Department to English at University of Delaware in 2006 and the Philip Lee Phillips Society Lecture at the Library of Congress in October 2006. I am very grateful to the many people involved in the organization of these invitations. They provided an opportunity to develop ideas and reminded me of the work yet to be done.

PHOTOGRAPHIC ACKNOWLEDGEMENTS

The author and publishers wish to express their thanks to the following sources of illustrative material and/or permission to reproduce it:

Archivo General de Indias, Seville: 4 (Est.145. Caj. 7. Leg. 8 Ramo 272. Mapas Mexico, 1); photos by the author: 3, 13, 23, 27, 28, 29, 30, 31, 42; Beinecke Library, New Haven, CT (Yale Collection of Western Americana, Beinecke Rare Book and Manuscript Library): 18, 19, 22, 25, 26, 32, 33, 34, 35, 36, 37, 38, 43, 44, 45, 46, 47, 49; Library of Congress, Washington, DC: 1, 5, 6, 14, 15, 20, 21, 41; photo courtesy of the Library of Congress, Washington, DC (Rare Books and Special Collections Division): 24; Public Record Office, London: 7 (C.O. 700 North American Colonies General no. 6(1)); Hudson's Bay Company Archives, Winnipeg: 8, 9 (E3/2 folios 106 verso-107); New York State Archives, Albany, NY: 11, 12; US Geological Survey Photographic Library, Reston, VA: 39, 40; photo courtesy of the US Geological Survey Photographic Library: 48.

INDEX

Numerals in *italics* indicate illustration numbers.